Praise for *Active Hope*

"Hope without action is naive optimism. Action without hope can fail because it doesn't believe in its goals. *Active Hope* is a compassionate and wise guide to bringing these forces together to understand who we are, what we need, and what we are capable of. It is also a brilliant guide to navigating the relationship between how we imagine ourselves and the world and what we can do in it."

— Rebecca Solnit, author of *Men Explain Things to Me* and *Hope in the Dark*

"The response to the profound crisis we are living through needs to be urgent, appropriate, and driven by possibilities. It must also be rooted in compassion and a focus on bringing us together in such a way that we can create anything we set our minds to. It is to this challenge that Joanna and Chris turn their minds here, and I for one am deeply grateful for the wisdom they offer us."

— Rob Hopkins, cofounder of the Transition Network and author of *The Transition Handbook: From Oil Dependency to Local Resilience*

"Words cannot fully express my gratitude for this book's appearance at this time. With ancient wisdom, deep compassion, and years of experience, Joanna and Chris guide us through this perilous time, not that we might succeed in saving the world, but that we might consciously choose to participate, no matter the outcome. This is the path of right action and right relationship, where joy and peace are available independent of what's going on around us. May we take this beautiful book into our hearts that we may each find the path with heart."

— Margaret J. Wheatley, author of *Perseverance*, *Leadership and the New Science*, and other books

"I loved this book. Its brilliance lies in the way it skillfully addresses and transforms the major limiting beliefs that hold people back from wholehearted social action, which at its heart requires that 'active hope' be sustained. It then gently, but inexorably, invites us to participate in the great adventure of our time of changing the world. Joanna and Chris have laid out a twenty-first-century spiritual path for deep and soul-filled social action. Read it and be nourished!"

— David Gershon, author of *Social Change 2.0: A Blueprint for Reinventing Our World* and codirector of Empowerment Institute's School for Transformative Social Change

"Our species is going through a collective Dark Night of the Soul. The authors call this the 'Great Turning,' and they provide ample nourishment in the form of concepts and practices that are sure to help us navigate our way. This book is a lucid, timely, practical, and very-much-needed gift. With the guidance offered, we learn how to stay true to our vocations so that neither fear nor despair nor uncertainty nor opposition defeats us. A welcome manual for navigating the Dark Night of our species. Most welcome!"

— Matthew Fox, author of *The Hidden Spirituality of Men* and *Christian Mystics*

"Joanna Macy is one of the great teachers of this age. What a delight, then, to read her latest work, written with medical doctor and addiction specialist Chris Johnstone. *Active Hope* helps us to gaze unflinchingly at the horrors that confront us, to honor the pain we feel for our world, and to allow the truth to strengthen rather than paralyze us. It's hard to think of a more important task than to empower us ordinary folk with tools that allow us to engage with the monumental challenges that confront us — tools that nourish our creativity and bolster our confidence and resilience. If you want to serve the living Earth but don't know how, if you feel overwhelmed by the magnitude of the task ahead, if you want to find the part that only you can play in the 'great work,' please read this book."

— John Seed, OAM, founder of the Rainforest Information Centre

"*Active Hope* is the Great Turning's New Testament. It is that rarity in the literature of self-help that unfolds as if organically in prose that, like the best sacred writing, induces the mind shift it encourages. Here are the unflinching diagnosis and prognosis and widely tested protocol for self-healing and lifesaving that can help a critical mass of us to recommit to learning, living, and acting effectively on behalf of Earth's beleaguered human and natural communities. *Active Hope* is laced with vivid analogies, anecdotes, and opportunities to envision the world we would wish to leave our children and grandchildren. I began reading it because I was asked to; I kept reading because I needed to. For all who are worried for the success of their work in the sustainability, environmental, new economy, or social justice movements, this book will be both a guide and a medicine."

— Ellen LaConte, author of *Life Rules: Nature's Blueprint for Surviving Economic and Environmental Collapse*

"*Active Hope* offers a way of living creatively in a time of great challenges, a time when humanity is threatened by ecological, social, and economic breakdown. Joanna and Chris show us how to frame our actions so that we can help contribute to the global movement for a better, more equitable, and ecologically balanced world. This is a juicy, creative, and powerful book."

— Maddy Harland, editor of *Permaculture* magazine,
www.permaculture.co.uk

"Active hope is something you choose, we're told in this powerful, inspiring book. Joanna Macy and Chris Johnstone offer not just an antidote to despair but a new lease on life, a way to do our part to heal our broken world. *Active Hope* is one of the most important works to appear in years and should be read by everyone young and old who cares about what is happening to our world. Pass it on and join the campaign to spread hope! It's never too late to make a difference."

— Rev. John Dear, lecturer, activist, and author of *Living Peace*,
The Questions of Jesus, *Transfiguration*, *Put Down Your Sword*,
The God of Peace, and *Seeds of Nonviolence*

"There are few guides in our world that are as trustworthy and brilliant as Joanna Macy! In *Active Hope*, she and Chris Johnstone give us not only some very good analysis but also good responses and practices — in a very readable and engaging style."

— Fr. Richard Rohr, OFM, Center for Action and Contemplation,
Albuquerque, New Mexico

"Before turning these pages, I was experiencing the total darkness of a midnight depression. I felt incapable of thinking through, much less taking, a necessary step in my life. Then I opened to the pages of Joanna and Chris's exercise 'The Bodhisattva Perspective.' Within minutes, I saw an otherwise inconceivable hope in the interconnected life I had chosen. It was a simple miracle of transformed consciousness. *Active Hope* is not just a book but a gateway to transformation."

— James W. Douglass, author of *JFK and the Unspeakable*

"Renowned Earth elder Joanna Macy has long exemplified what it means to live a life of spiritual activism and courageous compassion. Here, with her colleague Chris Johnstone, she offers the essential guidebook for everyone awakening to both the perils and potentials of our planetary moment. In this clearly written and compelling manual of cultural transformation, Joanna and Chris guide us to find hope where we might least have thought to look — within our own hearts and souls, and in our interdependence with all life — and then to boldly act on that hope as visionary artisans of life-enhancing cultures. To the future beings of the twenty-second century, *Active Hope* might turn out to be the most important book written in the twenty-first."

— Bill Plotkin, author of *Soulcraft* and *Nature and the Human Soul*

"Given the state of the world, how do we find a truly sane, effective, and life-affirming response? Joanna Macy, one of the true wisdom voices of our age, has devoted her life to helping people find their resilience, their creativity, and their passion while living with their eyes and hearts open. Her work is powerful, needed, and lifesaving. If you have despaired for our world, and if you love life, *Active Hope* will be for you an extraordinary blessing."

— John Robbins, author of *Diet for a New America* and *The Food Revolution*

"We belong to the Earth, and our precious living world is in crisis. More than any book I've read, *Active Hope* shows us the true dimensions of this crisis, and the way our heart and actions can be part of the Great Turning toward healing. Please read this book and share it with others — for your own awakening, for our children, and for our future."

— Tara Brach, PhD, author of *Radical Acceptance*

"This is a powerful book to read in these dark times. Written with wisdom and passion, it is about balanced compassion and hope, love and strength, and the will to make a difference. *Active Hope* is a brilliant guide to sanity and love."

— Roshi Joan Halifax, abbot of the Upaya Zen Center

ACTIVE
HOPE

Also by Joanna Macy

Books

Despair and Personal Power in the Nuclear Age

Dharma and Development: Religion as Resource in the Sarvodaya Self-Help Movement

Thinking Like a Mountain (with John Seed, Pat Fleming, and Arne Næss)

Mutual Causality in Buddhism and General Systems Theory

Rilke's Book of Hours (with Anita Barrows)

In Praise of Mortality: Selections from Rainer Maria Rilke's Duino Elegies and Sonnets to Orpheus (with Anita Barrows)

Widening Circles: A Memoir

A Year with Rilke (with Anita Barrows)

Pass It On: Five Stories That Can Change the World (with Norbert Gahbler)

Coming Back to Life: The Updated Guide to the Work That Reconnects (with Molly Young Brown)

A Wild Love for the World: Joanna Macy and the Work of Our Time (edited by Stephanie Kaza)

Rilke's Letter to a Young Poet (with Anita Barrows)

World as Lover, World as Self: Courage for Global Justice and Ecological Renewal

Audiovisual

The Work That Reconnects (DVD)

Also by Chris Johnstone

Books

Find Your Power: A Toolkit for Resilience and Positive Change

Seven Ways to Build Resilience: Strengthening Your Ability to Deal with Difficult Times

ACTIVE HOPE

How to Face the Mess We're in with Unexpected Resilience and Creative Power

JOANNA MACY & CHRIS JOHNSTONE

REVISED EDITION

New World Library
Novato, California

New World Library
14 Pamaron Way
Novato, California 94949

Interior illustrations on page 39 by Dori Midnight, on pages 150 and 152 by Dave Baines, and on pages 15, 16, 21, 42, and 100 by Carlotta Cataldi

Text design by Tona Pearce Myers
Cover design by Tracy Cunningham

Library of Congress Cataloging-in-Publication data is available.

First revised edition printing, June 2022
ISBN 978-1-60868-710-7
Ebook ISBN 978-1-60868-711-4
Printed in Canada on 100% postconsumer-waste recycled paper

New World Library is proud to be a Gold Certified Environmentally Responsible Publisher. Publisher certification awarded by Green Press Initiative.

10 9 8 7 6 5 4 3

This book is dedicated
to the flourishing of life
on this rare and precious Earth
and to the role each of us can play
in responding to our planetary emergency.

Contents

Acknowledgments

We begin with gratitude, remembering the words of Thich Nhat Hanh: "If you are a poet, you will see clearly that there is a cloud floating in this sheet of paper. Without a cloud, there will be no rain; without rain, the trees cannot grow; and without trees, we cannot make paper."[1] So it is with this book. Without all those who've played supporting roles, it would simply not be here. So our thanks extend to all who have helped, particularly:

The core team involved in its production. Early on in our book-writing journey, we were joined by two people who understood what we were setting out to do and who were with us in this purpose: our agent, Suresh Ariaratnam, and our editor, Jason Gardner. What a relief to have you alongside us still, as we bring out this revised tenth anniversary edition. Thank you. We're grateful to Dori Midnight for the wonderful spiral illustration in chapter 2, to Dave Baines for the two intriguing spirals of time in chapter 8, and to Carlotta Cataldi for the illustrations in chapters 1, 2, and 5. Thank you to all at New World Library for the different roles they've played, particularly Diana Rico (this edition) and Mimi Kusch (first edition) for copy-editing, Monique Muhlenkamp and Kim Corbin for publicity, Tona Pearce Myers for the interior design, Tracy Cunningham for the cover design, Danielle Galat for international editions, and Munro Magruder for marketing. We would like also to thank the many who have inspired us, particularly those we've quoted, such as Dr. A.T. Ariyaratne, Tom Atlee, John Seed, Rebecca Solnit, John Robbins, and the late Elise Boulding, Nelson Mandela, Arne Næss and Thich Nhat Hahn.

The many friends and family who have supported us as we were writing. Joanna's husband, Fran, supported us right from the beginning. We have felt him as a powerful ally, even after his death in 2009. Joanna particularly thanks her assistant Anne Symens-Bucher; her children, Peggy, Jack, and Christopher; and her grandchildren, Julien, Eliza, and Lydia. Chris's wife, Kirsty, encouraged him through writing ups and downs. Hamish Cormack interviewed Chris about each chapter and commented on early drafts. Our dear friend Kathleen Sullivan, who has known us both for decades, has been a friend of our writing too, and we thank her.

The countless colleagues in the Work That Reconnects, who have engaged with it as a practice, offered it in places all over the planet, and added their own distinctive contributions. As we take the essence of this approach to wider audiences, we are aware of the ways you have enriched it; our pleasure in your company is boundless.

The growing international team of *Active Hope*, which helps us bring *Active Hope* into the world, not just as a book, but also as a practice and stance in life. We include here translators and publishing teams behind editions in fourteen languages now and more in development. We appreciate the many who've engaged in *Active Hope* book groups and study-action circles. Thanks also to Elle Adams for website design at ActiveHope.info and the team at ActiveHope. Training, particularly Madeleine Young, for helping develop our YouTube channel and free online course. Lastly, if you've played a part in sharing *Active Hope* and passing it on, we thank you too. Welcome to our team.

Introduction

The year 2020 began with dead birds falling from smoke-filled skies as wildfires raged in Australia. It ended with a global death toll of nearly two million people killed by a pandemic that was still on the rise. Looking back, this was the year the Great Unraveling went mainstream.

Do you have the feeling that our world is falling apart? Or that our human civilization and planetary ecosystem are in dangerous decline? That's what we mean by *the Great Unraveling*, a term suggested by visionary economist and author David Korten when considering how future generations might talk about our time.[1] The Great Unraveling is more than just a few isolated disasters. As we'll explore, there's a larger pattern at work.

Canaries were once used in coal mines to provide warning of toxic gas. When the birds died from breathing invisible poisons in the air, miners knew they were in danger and needed to act quickly. The massive decline in global bird populations tells us something similar. Since 1970, nearly three billion birds have died in North America alone. Scientists have commented that this staggering loss suggests the very fabric of North America's ecosystem is unraveling.[2] Happening throughout our whole world, it is not just ecosystems that are unraveling, but our social systems too.

Over the past fifty years, each decade has become warmer. Extreme weather events, such as record-breaking heat waves, wildfires, droughts, floods, and storms, have become more common, a trend that will make future conditions much worse.[3] As global heating continues, the proportion of land becoming too hot to live on is set to increase from 0.8 percent in 2020 to 19 percent in 2070.[4] This could displace between one and three billion people.

Those who are younger will experience these changes most, with children born in 2020 between two and seven times more likely than their grandparents to suffer through floods, heat waves, droughts, wildfires, and crop failures.[5] The term *climate justice* draws attention to both this intergenerational injustice and the ways climate disruption disproportionately harms people living in low-income communities and countries.

Climate change isn't the only problem we face. Human population and consumption are increasing at the same time as essential resources, such as fresh water, fish stocks, and topsoil, are in decline. Extreme inequality is on the rise, with more and richer billionaires accumulating wealth in a world where hundreds of millions of people still starve. While the economic impact of the Covid-19 pandemic has left many feeling desperate about how they're going to manage, $2 trillion is spent globally each year on preparing for and engaging in warfare.[6]

Hope is often thought of as the feeling that things are going to get better. When facing the mess we're in, it is difficult for most of us to have that. Looking into the future, we can no longer take it for granted that the resources we depend on — food, fuel, and drinkable water — will be available. We can no longer even be certain that our civilization will survive or that conditions on our planet will remain hospitable for complex forms of life.

We are starting out by naming this bleak uncertainty as a pivotal psychological reality of our time. We may wonder whether we — our families, our civilization, and even our species — will make it. Yet because fears for our collective future are usually considered too uncomfortable to talk about, they tend to remain an unspoken presence at the back of our minds. We often hear comments such as "Don't go there, it is too depressing" and "Don't dwell on the negative." The problem with this approach is that it closes down our conversations and our thinking. How can we even begin to tackle the mess we're in if we consider it too depressing to think about or discuss? This blocked communication generates a peril even more deadly, for the greatest danger of our times is the deadening of our response.

Yet when we do face the mess — when we do let in the dreadful news of multiple tragedies unfolding in our world — we can feel overwhelmed. We may wonder whether we can do anything about it anyway.

So this is where we begin: by acknowledging that our times confront us with realities that are painful to face, difficult to take in, and confusing to live with. Don't be surprised if you find yourself feeling anxious, defeated, or in despair.

There's something else we'd like to bring in alongside this difficult starting point. It is a recognition that when we're at our most exasperated, we can sometimes surprise ourselves. We might discover strengths we never knew we had or experience degrees of aliveness we'd not even suspected were available to us. This is a time to reach out and find new allies, as well as to discard forms of thinking and behavior that have led us astray. In a process known as *adversity activated development*, our very act of facing the mess we're in can help us discover a more enlivening sense of what our lives are about, what we're here to do, and what we're truly capable of.

Do you hope this will happen for you? Or that you might play a role in helping this happen for others? If so, we invite you to join us in our journey. Together we will explore how we can access unexpected resilience and creative power, not just to face the mess we're in, but also to play our part in doing something about it.

WHAT IS ACTIVE HOPE?

Whatever situation we encounter, we can choose our response. When facing overwhelming challenges, we might feel that our actions don't count for much. Yet the kind of responses we make and the degree to which we believe they count are shaped by the way we think and feel about hope. Here's an example.

Jane cared deeply about the world and was horrified by what she saw happening. She regarded human beings as a lost cause, as so stuck in our destructive ways that she believed the complete wrecking of our world was inevitable. "What's the point of doing anything if it won't change what we're heading for?" she asked.

The word *hope* has two different meanings. The first, which we've touched on already, involves hopefulness, where our preferred outcome seems reasonably likely to happen. But if we require this kind of hope before we commit ourselves to an action, our response gets blocked in areas where we don't rate our chances well. This is what happened for Jane — she felt so hopeless she didn't see the point of even trying to change things.

The second meaning is about desire. When Jane was asked what she'd like to have happen in our world, without hesitation she described the future she hoped for, the kind of world she longed for so much it hurt. It is this kind of hope that starts our journey — knowing what we hope for and what we'd like, or love, to have happen. It is what we do with this hope that really makes the difference. Passive hope is about waiting for external agencies to bring about what we desire. Active Hope is about becoming active participants in the process of moving toward our hopes and, where we can, realizing them.

Active Hope is a practice. Like tai chi or gardening, it is something we *do* rather than *have*. It is a process we can apply to any situation, and it involves three key steps. First, we start from where we are by taking in a clear view of reality, acknowledging what we see and how we feel. Second, we identify what we hope for in terms of the direction we'd like things to move in or the values we'd like to see expressed. And third, we take steps to move ourselves or our situation in that direction.

Since Active Hope doesn't require our optimism, we can apply it even in areas where we feel hopeless. The guiding impetus is intention; we *choose* what we aim to bring about, act for, or express. Rather than weighing our chances and proceeding only when we feel hopeful, we focus on our intention and let it be our guide.

Building Our Capacity and Commitment

Most books addressing global issues focus on describing either the problems we face or the solutions needed. While we touch on both of these, our focus is on how we can strengthen our commitment and capacity to act, so that we can best play our part, whatever that may be, in the healing of our world.

Since we each look out onto a different corner of the planet and bring with us our own particular portfolio of interests, skills, and experiences, we are touched by different concerns and called to respond in different ways. The purpose of this book is to help you address the issues you're most moved by, guided by what's heartfelt and rings true to you.

When we become aware of an emergency and rise to the occasion, something powerful gets switched on inside us. We activate our sense of purpose and discover strengths we didn't even know we had. Being able to make a difference is powerfully energizing; it makes our existence feel more worthwhile. When we practice Active Hope, it is not about being dutiful or worthy so much as it is about stepping into a state of aliveness that transforms our lives and our world.

Three Stories of Our Time

In any great adventure, there are always obstacles in the way. The first hurdle is just to be aware that we, as a civilization and as a species, appear to be heading toward catastrophe. Even though there's more talk these days of the need to address our global crisis, most governments and mainstream businesses still pursue the same short-term priorities that have created the mess we face.

In chapter 1 we draw attention to the huge gap between the scale of the emergency and the size of the response, explaining this by exploring how our perceptions are shaped by the story we identify with. We describe three stories, or versions of reality, each serving as a lens through which we can see and understand what's going on.

We've already introduced our first story. We call it the Great Unraveling to name the progressive decline and collapse of countless interlocking elements in the social, ecological, hydrological, and atmospheric systems on which life depends. Climbing levels of hate crime, falling levels of trust, and increasing use of lies by political leaders all play a role.

When we step back and see this larger story as a whole, it is easier to recognize that our world isn't simply unraveling — it is being pulled apart. An active process is at work that's linked to a second story we call Business as Usual.

With Business as Usual, the defining assumption is that there is little need to change the way we live. Economic growth is regarded as essential for prosperity, and the central plot is about getting ahead, without any longer-range consideration about where this approach is taking us. Until 2020, this narrative so dominated mainstream culture that for many it seemed to be the only story on offer and accepted as just the way things are. Then came Covid-19. As disease, death, and disruption touched so many corners of our world, unraveling moved center stage. When there are years of upheaval, many people understandably yearn for a bounce back to how things used to be. Yet a return to Business as Usual, as we'll explore, just speeds up the unraveling.

The third story is embraced by those who believe a different way forward is possible and already underway. It includes those who stand firm to defend the social and ecological fabric of our world, engaging in activism for justice and the preservation of life. It also features those reinventing the way we do things, developing systems and structures that support the flourishing of life, from local currencies to solar cooperatives, from ecovillages to regenerative agriculture. This story encompasses both inner and outer dimensions of change, with shifts in our sense of who we are, who we want to be, and how we relate to one another and our living Earth.

The different threads of this third story combine and interact in a larger narrative arc: that of the epochal transition from the industrial growth society to a life-sustaining society. This far-reaching change is comparable in scope and magnitude to both the Agricultural Revolution ten thousand years ago and the Industrial Revolution less than three centuries back. We call this story the Great Turning.

When we look for them in the world around us, we can see elements from each of these stories at play. While unevenly distributed, all three are happening. There are hot spots of unraveling, as well as disturbing trends that will push it into greater prominence. There are also contexts where Business as Usual so dominates the landscape that it seems the only story visible. Set against these two, the Great Turning is often squeezed out of sight or deliberately underreported by mainstream media. We're going to be exploring a process that more clearly brings it into view. A great question to start this process is: "What happens through you?"

Stories happen through people; our choices and actions shape how they happen through us. If we shift focus from outcome to process, rather than asking, "Will it happen?," we might ask instead, "What are the steps that will help it happen?" We help the

Great Turning happen by turning up with an intention to play our part. We invite you to consider some of the different ways you are already doing that. What are you turning away from because you know it causes harm? What are you turning toward because it aligns more strongly with your values and hopes for our future? Noticing and appreciating ways the Great Turning is already happening through us, both personally and collectively, is a starting point we can build on. By itself, though, that isn't enough. We also need to apply Active Hope.

As Business as Usual becomes more difficult to sustain and the Great Unraveling advances, we'll face more severe consequences of the societal and ecological process of collapse that is already underway. What we don't know is what specific form it will take and at what pace. Some of the possible versions of how this might go are far more disturbing than others. Taking in the reality of what's happening in our world, what's the best we can hope for? And how can we be active in making that more likely or even possible?

While the pandemic has brought tragedy and loss for so many, it has also given examples of what can happen when we rise to the occasion in an emergency. You may have witnessed or been part of spontaneous networks of mutual aid springing up around the globe. People stopped flying, factories spewed less smoke, the air became cleaner, and carbon emissions fell. Before these things happened, any of them might have been dismissed as unrealistic hopes. It raises the question of what might be possible if we, collectively, really set our hearts and minds to making the changes that are needed. To help this happen in the best possible way, we need to train ourselves.

The Spiral of the Work That Reconnects

The journey that begins in chapter 2 and that continues throughout the book is based on a transformative and strengthening process we have offered in workshops for decades. Initially developed by

Joanna in the late 1970s, it evolved and spread, with the vital contribution of a growing number of colleagues. It has been used on every continent (except Antarctica perhaps), has been conducted in many different languages, and has involved hundreds of thousands of people of different faiths, backgrounds, and age groups. Because this approach helps us restore our sense of connection with the web of life and with one another, it has come to be known as the Work That Reconnects.[7] By helping us to develop our inner resources and our outer community, the Work That Reconnects strengthens our capacity to face disturbing information and respond with unexpected resilience. In our experience of this work, again and again we've seen energy and commitment mobilized as people rise to their role in the Great Turning.

We've written this book so that you can experience the transformative power of the Work That Reconnects and draw on it to expand your capacity to respond creatively to the crises of our time. The chapters ahead guide you through the four stages of the spiral this work moves through: coming from gratitude, honoring our pain for the world, seeing with new eyes, and going forth (see figure 1). The journey through these stages has a strengthening effect that deepens with each fresh experience.

While rich rewards can be reaped when journeying alone, the benefits of the Work That Reconnects grow quickly with company. We encourage you to seek others with whom to read this book or share notes along the way. Some of the practices are also described in the free online course at ActiveHope.Training. Bringing our concerns into the open is a key part of facing the mess we are in, though for reasons we will explore, fear often prevents this type of sharing. We will examine what makes it so difficult to talk about our planetary crisis and provide tools that support us in having the empowering conversations our times call for.

We encourage you to gain familiarity with the tools we describe

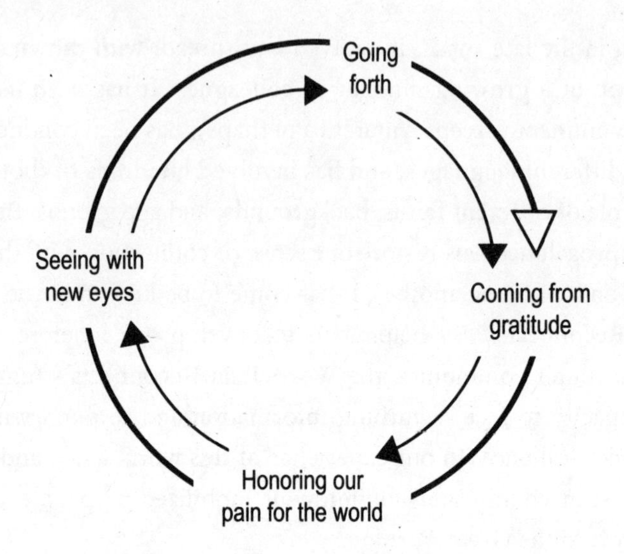

Figure 1. The spiral of the Work That Reconnects.

by trying them out. Scattered throughout the book are "Try This" exercises inviting you to experience practices we find valuable for personal use, in partnered conversations, and in groups.

WHAT WE BRING

At the heart of this book is a collaborative model of power based on appreciating how much more we can achieve working together than as separate individuals. The story of our coauthorship is a fine example. The seedling idea sprouted out of a conversation about lessons we, Joanna and Chris, had learned from our experience of the Work That Reconnects. What surprised and excited us both was how often, in the many hours of talking that followed, insights would surface that neither of us had received before. While the core framework, concepts, and practices of the Work That Reconnects are well tested, we have been able to enrich, hone, and add to them in ways that bring together a great deal of material not published elsewhere.

There is an old saying that two eyes are better than one, since out of two different perspectives comes the depth of three-dimensional vision. As coauthors we come from different backgrounds, live on different continents, and draw from different sources, all of which has contributed to the rich synergy we have experienced and expressed through our writing.

Joanna, a practitioner of Buddhism, general systems theory, and deep ecology, is the root teacher of the Work That Reconnects. In her early nineties and an activist for more than six decades, she lives in Berkeley, California, close to her children and grandchildren. She has authored or coauthored eighteen books, including four cotranslations of the German lyric poet Rainer Maria Rilke.

Chris is a medical doctor, coach, and trainer who has specialized in the psychology of resilience, behavior change, and recovery from addiction. Living in the United Kingdom, he offers online resilience courses that have engaged people from more than sixty countries. An activist since his teenage years and now entering his sixties, he has taught and written about the psychology of sustainability for more than thirty years.

The two of us met in 1989 at a weeklong training led by Joanna in Scotland. Called the Power of Our Deep Ecology, it was a life-changing event for Chris. We have worked together many times since. This book describes the work we share and cherish. It is offered, not as a blueprint solution to our problems, but as a set of practices and insights to draw strength from and as a mythic journey to be transformed by.

The author and activist Rebecca Solnit writes: "An emergency is a separation from the familiar, a sudden emergence into a new atmosphere, one that often demands we ourselves rise to the occasion."[8] When we face the mess we're in, we realize that Business as

Usual can't go on. What helps us rise to the occasion is experiencing our rootedness in something much larger than ourselves. The poet Rabindranath Tagore expressed this idea in these words: "The same stream of life that runs through my veins night and day runs through the world."[9] This is the stream we are following. It points us toward a way of life that enriches our world. Our foundation and our focus is the practice of Active Hope. When we face the mess we're in by engaging in this, our lives become enriched as well.

PART ONE

The Great Turning

CHAPTER ONE

Three Stories of Our Time

In the center of the room, three large pieces of paper lay on the floor. "That's where I lived for most of my early adult life," said Elly, pointing to the sheet nearest her. On it were the words "Business as Usual." "Then I spent the next four years in that one, the Great Unraveling, as I became more aware of the mess we're in and felt totally overwhelmed. It has only been in the last year or two that I've come to recognize this third story, the Great Turning, and find my place in it." We were at a workshop exploring Active Hope (see figure 2). The circle of people around the room nodded as they too reflected on their journeys of becoming aware of what we face and activated to respond.

Figure 2. We can look at the world from different stories.

From what story, or stories, do you look at the world? How has that changed over the course of your life? We're using the term *story* here to refer to the way our minds make sense of our world by placing aspects of what's happening within a larger narrative. Psychologists might call this our *metanarrative* — it is the larger pattern we see expressed through many contributing elements and events. In this chapter, we look more deeply at the three stories mentioned in the introduction: the Great Unraveling, Business as Usual, and the Great Turning. Recognizing that we can choose the story we live from can be liberating; finding a good story to take part in adds to our sense of purpose and aliveness. We will explore how each of these stories shapes our response to global crises.

THE GREAT UNRAVELING

How serious do you think the problems are that we face as a human society? If you were to rate your response on a scale of 0 to 10, where 0 is no problem at all and 10 represents catastrophe (see figure 3), what number would you choose?

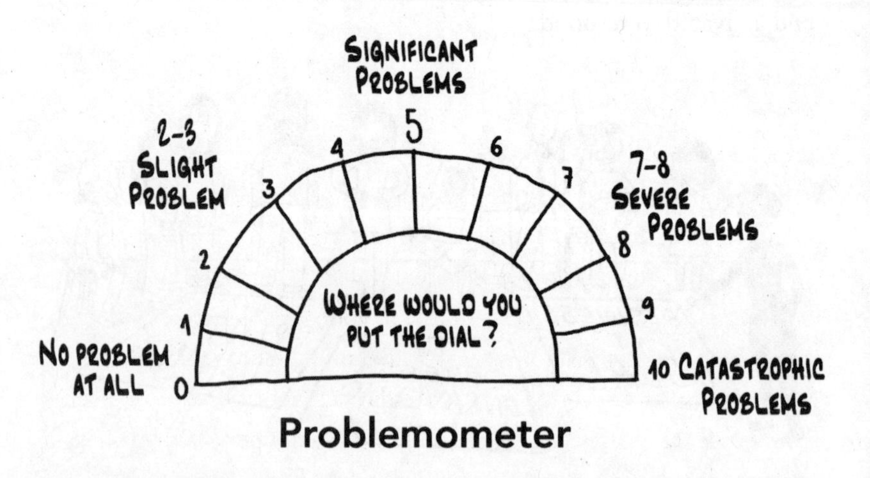

Figure 3. How severe are the problems we face?

If you've given any points at all, you're recognizing that something isn't right. If your score is more than 7, you're sensing we're in danger. What is it that concerns or alarms you? Naming your concerns brings them into focus. As the first step in addressing any issue is simply to notice it, this is the start of our journey.

Try This: Naming Your Concerns

What words seem to naturally follow these sentence starters for you? You can think them to yourself, or put them in writing, or try it with a partner, taking turns to speak and listen. Whenever you're not sure what to say, just speak out the beginning of the sentence and see what comes — it may be different each time you do this.

- When I consider our world situation, some concerns I have include . . .
- What particularly troubles me is . . .

When you begin mapping out the range of threats we face, what you're describing is the Great Unraveling. It is useful to give a name to this larger pattern of worsening conditions; though it is seldom talked about, it is something most people are aware of. For example, in 2017 an international poll of eighteen thousand people found that 67 percent thought the world was getting worse.[1] In 2021, an international poll of ten thousand young people revealed that 75 percent thought the future was frightening.[2] But what happens if you bring your concerns up in everyday conversation? Most likely you'll quickly receive strong messages that such open sharing of disturbing information is unwelcome. You might find yourself caught up in an argument. You risk being labeled as depressing company.

Even though there are cultural resistances in the way, bringing our fears into the open is a necessary step toward exploring how we can respond to them. Reviewing some of the conditions that concern us, the following stand out. Our naming them here is to acknowledge some of those we face rather than to attempt a more complete overview.

- The worsening climate crisis
- The threat of war and the cost of warfare
- Pollution and destruction of the natural world
- Widening of extreme inequality
- Dismantling of democracy
- Fears for how we'll meet our basic needs
- The prospect of societal collapse
- Misinformation and lies blocking awareness of what's really going on

The Great Unraveling is unevenly distributed and experienced differently, depending on where you are in the world and what conditions you face. You might start your own list with a different set of concerns, and you may have other issues you would add, as we do. Seeing a range of threats as connected elements within a larger story can help us identify ways they influence one another.

The term *unraveling* describes a process rather than an event. For some communities, the experience of having their world torn apart isn't a new thing: indigenous peoples have had their ancestral lands stolen; invading colonizers, through wealth grabs over centuries, have set in place a vastly unequal world order designed to serve the interests of a privileged minority. Nearly half the world's population lives on less than $5.50 a day,[3] and more than nine hundred million people were so short of food in 2021 they suffered from undernutrition.[4] One of the biggest contributors to unraveling worldwide is injustice; this interacts with, and amplifies, the harmful impacts of climate change, war, habitat decline, and pandemics.

Set against this background, the Great Unraveling is a process that is ongoing and, disturbingly, still gathering momentum. We can see this in the vicious cycles of worsening climate change. For example, forests play a protective role by absorbing carbon dioxide, but as woodlands are chopped down, we lose that crucial process. Large tropical trees are at additional risk because when warmer air dries out the soil, the ground can no longer support them. A global temperature increase of more than 7.2°F (4°C) could be enough to kill much of the Amazon rain forest, the "lungs of the planet."[5] If this happened, not only would we lose the rain forest's cooling effect, but the greenhouse gases released from rotting or burning trees would further increase warming. The term *runaway climate change* describes this dangerous situation, in which the consequences of warming cause more warming to occur (see figure 4).

As land and sea surfaces absorb more of the sun's warmth than ice cover does, the melting of ice sheets forms part of another vicious cycle. The more the ice melts, the less it reflects the sun's heat, and the warmer temperatures get, leading to further ice melting. Most of the world's major cities developed as ports bordering the sea or major rivers, and more than 630 million people live less than thirty-three feet above sea level. As the ice sheets in Greenland and West Antarctica continue melting, rising water levels will flood London, New York, Miami, Mumbai, Calcutta, Sydney, Shanghai, Jakarta, Tokyo, and many other major cities.[6]

When we recognize the Great Unraveling, we're not just acknowledging disasters that have already happened — we're bringing into awareness trends already firmly established that will make future conditions worse.

A central mechanism driving the unraveling process is the system dynamic of *overshoot and collapse*. Whether with individual

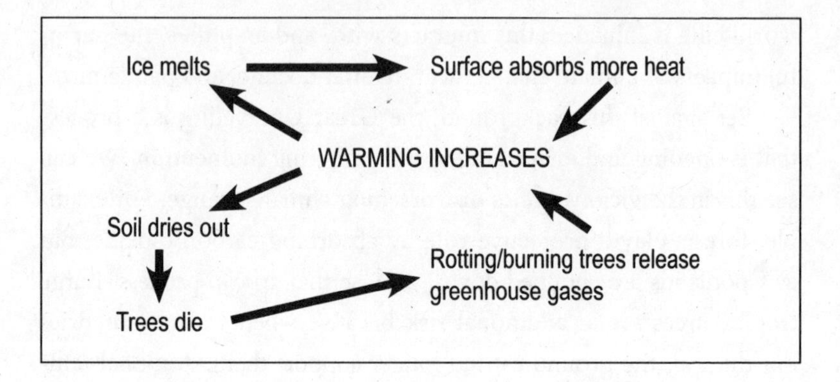

Figure 4. Amplifying loops in runaway climate change.

people, organizations, or ecosystems, pushing something too far beyond its coping limit takes it toward a state of breakdown. For example, when someone experiences severe stress for prolonged periods, their inner reserves can get so depleted that they reach a point of collapse. Something similar happens in agricultural systems when overfarming of land causes the collapse of soil fertility. As a result, each year, an area of once-fertile land the size of England gets turned into desert.[7]

Unraveling is a useful term because it implies a process that progressively lessens and loosens a system's coherence, memory, and functionality until it falls apart. There are different stages along this route of decline. At first there might be warning signs, as edges become frayed. Then capacities begin to disappear as significant elements and functions are lost. Eventually a breaking point is reached when the system, no longer able to hold itself together, collapses. We can see this sequence being played out with so many aspects of our world — agricultural systems, fisheries, natural habitats, communities living in degraded landscapes, organizations struggling with worsening conditions, and even whole countries.

In an interconnected world, collapse in one area has ripple effects that influence other systems. In the *Future Earth Risks*

Perceptions Report 2020, a group of more than two hundred scientists from fifty-four countries warned that the colliding impacts of climate change, extreme weather events, ecosystem decline, food crises, and freshwater shortages act together and amplify one another, setting us on a course toward a global systemic collapse.[8] Recognizing the psychological impact of living with the prospect of such catastrophic collapse, Jem Bendell and Rupert Read, in their book *Deep Adaptation*, write:

> Not only is it difficult to allow this outlook into one's awareness, it is difficult to live with it because to anticipate societal collapse means we feel personally vulnerable as well as afraid for the future of people dear to us.[9]

Acknowledging the scale of the crisis we face, you might hope for a mass mobilization that draws together governments, organizations, communities, and people everywhere to rise to the occasion of our planetary emergency. Yet do we see that? We can have a second assessment scale that this time asks how well developed you think our collective response is to these problems (see figure 5). In

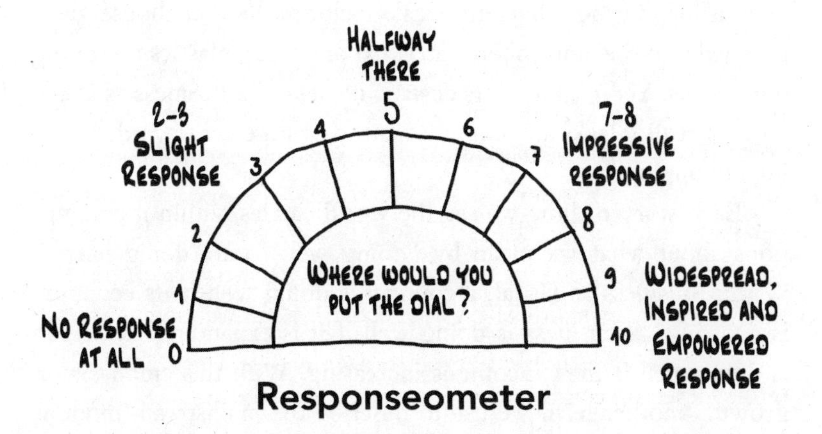

Figure 5. How well developed is our collective response?

this case, 0 is no response at all, and 10 is as good as it could possibly get, with a widespread, well-developed, and inspiring international response. How would you rate our current collective response?

Business as Usual

Did your ratings on the two scales show a mismatch between the severity of our problems and the size of our collective response? If you, like us, gave the problems a much higher score, you'll likely recognize the power of our second story, Business as Usual. The core assumption here is that things aren't too bad and that we can carry on our business the way we usually do. When looking at the world from this perspective, periodic disasters are seen as only temporary interruptions.

With Business as Usual, we're referring to the mainstream mode of industrialized countries and the ways of doing things considered normal within them. For example, it is common for people today to consume much higher levels of energy and materials than previous generations and to produce vastly more waste. The rubbish generated each year in the United States could fill a convoy of garbage trucks long enough to go round the world nine times.[10] The less visible but more harmful waste includes the greenhouse gases pumped into the atmosphere each year and microplastics poisoning our oceans. Yet in spite of its destructiveness, the Business as Usual view of reality lacks any commitment to or urgency toward changing our approach.

Each story of how we see the world carries within it assumptions about what we mean by "doing well" and "doing harm." Within Business as Usual, a country is doing well if its economy is growing. A business is doing well if it is expanding. A person is doing well if their income is increasing. With this emphasis on growth, another term we use to describe the mainstream mode is "the industrial growth society." For industry to grow, more sales are needed. That means encouraging us to buy and consume more than we already do. This drives the dynamic of overshoot and collapse, as the constant pressure to produce more things requires the extraction of resources, which gobbles up landscapes and then poisons our world with the waste generated.

The harmful norms of Business as Usual are held in place by a set of assumptions about how the world operates. These core assumptions include:

- Economic growth is essential for prosperity.
- Nature is a resource to be used for human purposes.
- Promoting consumption is good for the economy.
- Life is unequal, and some lives matter more than others.
- The problems of other peoples, nations, and species are not our concern.
- There's no point in worrying about the distant future, as we'll be dead by then.
- What we do doesn't make any difference; we can't change the world.

The Dilemma of the Double Reality

The stories of Business as Usual and the Great Unraveling offer starkly contrasting accounts of the state of our world. They are two different realities coexisting in the same time and space. You probably know people who live a different story than you. You may also be moving between stories yourself. It's possible to spend part of a day in our own Business as Usual mode, making plans for a future we assume will be much like today. Then something triggers an awareness of the mess we're in, and we recognize in our hearts and minds the crash that lies ahead or that is being experienced already.

When people first become aware of the extent of the unraveling, it comes as quite a shock. The mainstream mode has been to avoid looking at issues if they're too disturbing, so when we fully open our eyes to the horror, we can feel unprepared, overwhelmed, and defeated. Many people we've spoken with have described flip-flopping between these different stories, as though caught in a double reality. Living with these two opposing versions of how things are going presents us with an agonizing dilemma: to face the mess we're in is scary and depressing, yet to turn away and not face it can leave us feeling that we're living a lie, being complicit with a social order that is destroying our world. When we look at the worst, we may also experience a collapse in our belief that we can make any difference. When we avoid looking, we don't see the need to even try.

In the depths of a crisis, there can be moments when something that wasn't so easy to see becomes clearer. We live at a unique time in human history, when our collective experience of one traumatic process — in this case, the Covid-19 pandemic — can play a role in helping us recognize and see a way out of this dilemma of the double reality.

A Teachable Moment in Recent History

In health psychology, one of the most productive times for a conversation about change is just after someone has experienced the harmful consequences of their behavior. For a heavy smoker, when the distress of a chest infection serves as a wake-up call to address their habit, we can think of this as a teachable moment. At the United Nations General Assembly in September 2020, UN Secretary-General António Guterres suggested that the shared global tragedy of the pandemic might offer something similar for our civilization:

> The pandemic is a crisis unlike any we have ever seen. But it is also the kind of crisis that we will see in different forms again and again. COVID-19 is not only a wake-up call, it is a dress rehearsal for the world of challenges to come.[11]

For a teachable moment to happen, we need to draw out the lessons a situation, even a tragic one, might offer. If the pandemic can act as a wake-up call for anything, perhaps at the top of the list is the danger of continuing with Business as Usual when that makes the current tragedy so very much worse.

With less than 7 percent of the world's population between them, Brazil and the United States suffered nearly a third of the global death toll from Covid-19 in 2020.[12] Both countries were led by presidents who approached the pandemic and climate change in similar ways, by downplaying concerns and delaying any departure from Business as Usual. We don't have to continue making the same mistake.

When disasters happen, we can easily feel powerless. A second learning point is how personal change and policy change can act together in complementary ways to bring about rapid transformation. As the pandemic spread to all corners of the globe, it required both personal and policy shifts to flatten the curve of rising infections and bring down the death rate. If we hadn't acted together, with huge shifts at the level of individuals, families, communities, organizations, governments, and international bodies, it is estimated that there would have been twenty times as many deaths worldwide in 2020 and many more after that.[13] By putting the brakes on Business as Usual, we stopped a disaster from developing into a catastrophe. Similarly, in the realm of climate change, we have another curve to flatten and then bring down with our greenhouse gas emissions. As with the pandemic, by acting sooner and more sharply, we can reduce the chances of the very worst outcomes.

Never before have we experienced a threat that touched the lives of such a high proportion of our world's population in such a short period of time. It has exposed our shared vulnerability. António Guterres named a third lesson so relevant to this when he said: "In an interconnected world, it is time to recognize a simple truth: solidarity is self-interest. If we fail to grasp that fact, everyone loses."[14]

These three learning points — that continuing with Business as Usual takes us toward catastrophe, that we need to make rapid shifts at every level to avert this, and that solidarity is self-interest — form the core elements of our third story.

The Third Story: The Great Turning

While the responseometer of our collective mobilization doesn't yet show the high degree of universal engagement needed to address our planetary emergency, if you look for them, you can see examples of impressive steps toward what is required. In every country, in all walks of life, people are turning up with an intention to play their part. They are turning away from behaviors and ways of doing things that cause harm. They are turning toward ways of doing and thinking and being that support the flourishing of life. This is the Great Turning — and you are likely part of it.

Try This: Reflect on How the Great Turning Is Already Happening through You

We invite you to contemplate these questions:

- Where and how are you turning up?
- What are you turning away from?
- What are you turning toward?

Just as giving a name to the larger story of unraveling helps us recognize its horrific proportions, so too does viewing actions on behalf of life as central to a larger story help us see their power.

Each time you take a step, rather than dismissing it with the thought "That won't achieve much," consider instead how a grander narrative might express itself through the actions we take. It is not the first time that radical transformations have happened in our culture.

In the Agricultural Revolution of ten thousand years ago, the domestication of plants and animals led to a profound shift in the way people lived. In the Industrial Revolution that began just a few hundred years ago, a similar dramatic transition took place. These weren't just changes in the small details of people's lives. The whole basis of society was transformed each time, including people's relationships with one another and with the Earth.

Right now a shift of comparable scope and magnitude is occurring. It's been called the Ecological Revolution, the Sustainability Revolution, even the Necessary Revolution. This is our third story: we call it the Great Turning and see it as the essential adventure of our time. It involves the transition from a doomed economy of industrial growth to a life-sustaining society committed to the recovery of our world. In the early stages of major transitions, the initial activity might seem to exist only at the fringes. Yet we've already moved far beyond that, with many aspects of this transition already well underway. When their time comes, ideas and behaviors become contagious: the more people pass on inspiring perspectives, the more these perspectives catch on. At a certain point, the balance tips, and we reach critical mass. Viewpoints and practices that were once on the margins become the new mainstream.

In the story of the Great Turning, what's catching on is commitment to act for the sake of life on Earth as well as the vision, courage, and solidarity to do so. Social and technical innovations converge, mobilizing people's energy, attention, creativity, and determination, in what Paul Hawken describes as "the largest social movement in history." In his book *Blessed Unrest*, he writes, "I soon realized that my initial estimate of 100,000 organizations was off by at least a factor of ten, and I now believe there are over one — and maybe even two — million organizations working towards ecological sustainability and social justice."[15]

Don't be surprised if you haven't read about this epic transition in major newspapers or seen it reported in other mainstream media. Their focus is usually trained on sudden, discrete events they can point their cameras at. Cultural shifts happen on a different level; they come into view only when we step back far enough to see a bigger picture changing over time. If we look too close up when enlarging a digital image, we lose sight of the picture and just see pixels. Similarly, a newspaper photograph viewed through a magnifying glass appears only as tiny dots. When it seems as if our lives and choices are like those dots or pixels, it can be difficult to recognize their contribution to a bigger picture of change. We might need to train ourselves to see the larger pattern and recognize how the story of the Great Turning is happening in our time. Once seen, it becomes easier to notice it. As we name it and live it, this story becomes more real and familiar to us.

As an aid to appreciating the ways you may already be part of this story, we identify three dimensions of the Great Turning. They are mutually reinforcing and equally necessary. For convenience, we've labeled them as first, second, and third dimensions, but that is not to suggest any order of sequence or importance. We can start at any point and see its natural relevance to the others. It is for each of us to follow our own sense of rightness about where and how we act.

The First Dimension: Holding Actions

Holding actions aim to hold back and slow down the damage being caused by the political economy of Business as Usual. The goal is to protect what is left of our natural life-support systems, rescuing what we can of our biodiversity, clean air and water, forests, and topsoil. Holding actions also counter the unraveling of our social fabric, involving caring for those who have been damaged and safeguarding communities against exploitation, war, starvation, and injustice. Holding actions defend our shared existence and the integrity of life on this, our planet home.

This dimension includes raising awareness of the damage being done by gathering evidence of and documenting the environmental, social, and health impacts of industrial growth. We need the work of scientists, campaigners, and journalists to reveal the links between pollution and rising childhood cancers; fossil fuel consumption and climate chaos; and the availability of cheap products and sweatshop working conditions. Unless we see these connections clearly made, it is too easy to go on unconsciously contributing to the unraveling of our world. We become part of the story of the Great Turning when we increase our awareness, seek to learn more, and alert others to the issues we all face.

There are many ways we can act. We can choose to remove our support for behaviors and products we know to be part of the problem. Joining with others, we can add to the strength of campaigns, petitions, boycotts, rallies, legal proceedings, direct actions, and other forms of protest against practices that threaten our world. While holding actions can be frustrating when met with slow progress or defeat, they have also led to important victories. Through determined and sustained activism, acid rain is much reduced as a threat to forests and lakes, while the hole in the ozone layer has shrunk.[16] In the United Kingdom, fracking has been brought to a halt, while in Canada, the United States, Poland, Australia, and the Ukraine, areas of old-growth forests have been protected.

Holding actions are essential: they save lives, they save species and ecosystems, and they save some of the gene pool for future generations. But by themselves, they are not enough to create the Great Turning. For every acre of forest protected, many others are lost to logging or clearance. For every species brought back from the brink, many others are lost to extinction. Vital as protest is, relying on it as a sole avenue of change can leave us battle weary or disillusioned. Along with stopping the damage, we need to replace or transform the systems that cause the harm. This is the work of the second dimension.

The Second Dimension: Life-Sustaining Systems and Practices

If you look for it, you can find evidence that our civilization is being reinvented all around us. Previously accepted approaches to healthcare, business, education, agriculture, transportation, communication, psychology, economics, and so many other areas are being questioned and transformed. This is the second strand of the Great Turning, and it involves a rethinking of the way we do things, as well as creative redesigning of the structures and systems of our society.

In September 2021, Harvard University announced it would no longer invest its considerable endowment fund (of more than $41 billion) in the fossil fuel industry. Harvard's move to "decarbonize" its funds followed the example of 1,337 other institutions divesting more than $14 trillion of assets.[17] By steering investment away from industries that cause harm, the divestment movement is rewriting the rules of finance. Looking also at what we might turn toward, new types of banks, like Triodos Bank, operate on the principle of *triple return*. In this model, investments bring not only financial return but also social and environmental benefits. The more people put their savings into this kind of investment, the more funds become available for enterprises that aim for greater benefits than just making money. This in turn fuels the development of a new economic sector based on the triple bottom line. These investments have proved to be remarkably stable at a time of economic turbulence, putting ethical banking in a strong financial position.

One area benefiting from such investment is the agricultural sector, which has seen a swing to environmentally and socially responsible practices. Concerned about the toxic effects of pesticides and other chemicals used in industrial farming, large numbers of people have switched to buying and eating organic produce. Fair-trade initiatives improve the working conditions of producers, while community-supported agriculture (CSA) and farmers markets cut food miles by increasing the availability of local produce. In some sectors, like green building, design principles that were considered to be on the fringe a few years ago are now finding widespread acceptance. In these and other areas, strong green shoots are sprouting, as new organizational systems grow out of the visionary question "Is there a better way to do things — one that brings benefits rather than causing harm?"

When we support and participate in these emerging strands of a life-sustaining culture, we become part of the Great Turning. Through our choices about how to travel, where to shop, what to buy, and how to save, we shape the development of this new economy. Social enterprises, microenergy projects, community teach-ins, regenerative agriculture, and ethical financial systems all contribute to the emerging ecosystem of a life-sustaining society. By themselves, however, they are not enough. These new structures won't take root and survive without deeply ingrained values to sustain them. This is the work of the third dimension of the Great Turning.

The Third Dimension: Shift in Consciousness

What inspires people to embark on projects or support campaigns that are not of immediate personal benefit? At the core of our consciousness is a wellspring of caring and compassion. This aspect of ourselves — which we might think of as our *connected self* — can be nurtured and developed. We can deepen our sense of belonging to the world. Like trees extending their root system, we can grow in connection, thus allowing ourselves to draw from a deeper pool of strength, accessing the courage and intelligence we so greatly need right now.

If we surrendered
to Earth's intelligence
we could rise up rooted, like trees.

— RAINER MARIA RILKE

This dimension of the Great Turning arises from shifts taking place in our hearts, our minds, and our views of reality. It involves utilizing insights and practices that resonate with venerable spiritual traditions, while in alignment with revolutionary new understandings from science.

A significant event in this part of the story is the Apollo 8 spaceflight of December 1968. Because of this mission to the moon and the photos it produced, humanity had its first sighting of Earth as a whole. Twenty years earlier, the astronomer Sir Fred Hoyle had said, "Once a photograph of the Earth taken from the outside is available, a new idea as powerful as any in history will be let loose."[18] Bill Anders, the astronaut who took those first photos, commented, "We came all this way to explore the moon, and the most important thing is that we discovered the Earth."[19]

We are among the first in human history to have had this remarkable view. It came at the same time as the development in science of a radical new understanding of how our world works. Looking at our planet as a whole, the central insight in Gaia theory is that the Earth functions as a self-regulating living system.

During the past fifty-plus years, those first Earth photos, along with Gaia theory and environmental challenges, have provoked the emergence of a new way of thinking about ourselves. No longer just citizens of this country or that, there is a shift in consciousness toward a deeper collective identity. As many indigenous traditions have taught for millennia, we are all part of the Earth.

With this evolutionary jump comes a beautiful convergence of two areas previously thought to clash: science and spirituality. The awareness of a deeper unity connecting us lies at the heart of almost all spiritual traditions; insights from modern science point in a similar direction. We live at a time when a new view of reality is emerging, where spiritual insight and scientific discovery both contribute to our understanding of ourselves as intimately interwoven with our world.

Recognizing ourselves as a planet people, we cast aside the false hierarchy of value that sets one group above another, whether based on gender, race, ethnicity, class, sexual orientation, or other division. At the same time, if we've grown up in a society that allocates privilege according to a ladder of importance that has rich White men at the top, some of that type of thinking may still have a grip on us, even if we're not aware of it. A key aspect of the shift in consciousness is decolonization, where we're on the lookout for internalized expressions of a colonial mindset that we can acknowledge and address.

We take part in this third dimension of the Great Turning when we pay attention to the inner frontier of change, to the personal and spiritual development that enhances our capacity and desire to act for our world. By strengthening our compassion, we give fuel to our courage and determination. By refreshing our sense of belonging in the world, we widen the web of relationships that nourish us and protect us from burnout. In the past, changing the self and changing the world were often regarded as separate endeavors and viewed in either-or terms. But in the story of the Great Turning, they are recognized as mutually reinforcing and essential to each other (see figure 6).

Active Hope and the Story of Our Lives

Future generations will look back at the time we are living in now. The kind of future they look from and the story they tell about our

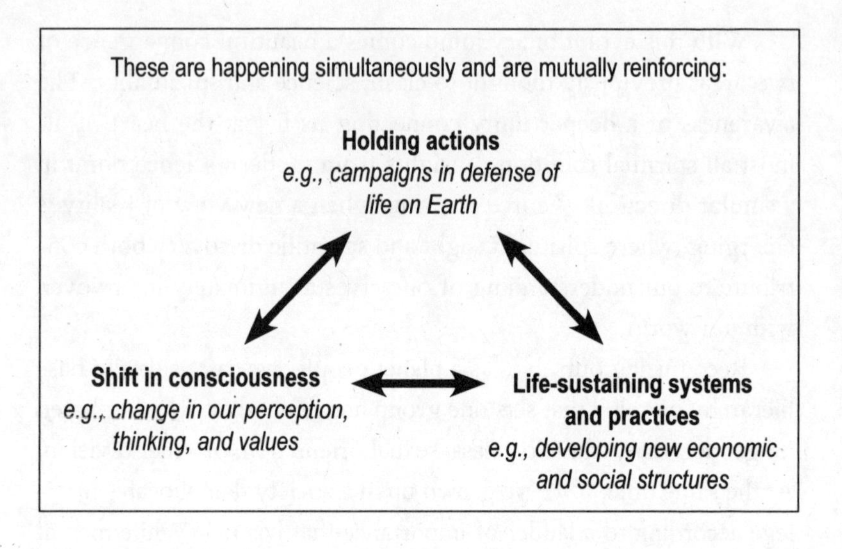

Figure 6. The three dimensions of the Great Turning.

period will be shaped by choices we make in our lifetimes. The most telling choice of all may well be the story we live from and see ourselves participating in. It sets the context of our lives in a way that influences all our other decisions.

In choosing our story, not only do we cast our vote of influence over the kind of world that future generations will inherit, but we also affect our own lives in the here and now. When we find a good story and fully give ourselves to it, that story can act through us, breathing new life into everything we do. Then a sense of deeper purpose adds to the meaning and momentum of our lives. A great story and a satisfying life share a vital element: a compelling plot that moves toward meaningful goals, where what is at stake is far larger than our personal gains and losses. The Great Turning is such a story.

CHAPTER TWO

Trusting the Spiral

Active Hope is not wishful thinking.
Active Hope is not waiting to be rescued
by the Lone Ranger or some savior.
Active Hope is waking up to the beauty of life
on whose behalf we can act.
We belong to this world.
The web of life is calling us forth at this time.
We've come a long way and are here to play our part.
With Active Hope we realize that there are adventures in store,
strengths to discover, and comrades to link arms with.
Active Hope is a readiness to engage.
Active Hope is a readiness to discover the strengths
in ourselves and in others,
a readiness to discover the reasons for hope
and the occasions for love.
A readiness to discover the size and strength of our hearts,
our quickness of mind, our steadiness of purpose,
our own authority, our love for life,
the liveliness of our curiosity,
the unsuspected deep well of patience and diligence,
the keenness of our senses, and our capacity to lead.
None of these can be discovered in an armchair or without risk.

The Great Turning is a story of Active Hope. To play our best part, we need to counter the voices that say we're not up to the task, that we're not good enough, strong enough, or wise enough to make any difference. If we fear that the mess we're in is too awful to look at or that we won't be able to cope with the distress it brings up, we need to find a way through that fear. This chapter describes three threads we can follow that help us stand tall and not shrink away when facing the immensity of what's happening to our world. These threads can be woven into any situation to reinforce and strengthen our capacity to respond. We shall therefore return to them often in the pages ahead.

THE THREAD OF ADVENTURE

The first thread is the narrative structure of adventure stories. These often begin by introducing an ominous threat that seems way beyond what the main characters are capable of dealing with. If you ever feel the odds are stacked against you and doubt whether you're up to the challenge, it can be useful to cast yourself as a character in the story of the Great Turning. The time-honored tradition of protagonists in this genre is that heroes and heroines almost always start out seeming distinctly underpowered.

What makes the story an adventure is that the central characters are not put off by the challenge they face. Instead, their tale sets them off on a quest in search of the allies, tools, and wisdom they need to improve their chances of success. We can think of ourselves as being on a similar journey; part of the adventure of the Great Turning involves seeking the company, sources of support, tools, and insights that can help us.

What starts us off is seeing what's at stake and feeling called to play our part. Then we just follow the thread of the adventure, developing capacities along the way and discovering hidden strengths that *reveal themselves only when needed.* When things are bumpy or bleak, we can remind ourselves that this is how these stories often go.

There may be times when all feels lost. That too can be part of the story. Our choices at such moments can make a crucial difference.

The Thread of Active Hope

Any situation we face can develop in a range of different ways — some much better, others much worse. Active Hope involves identifying the outcomes we hope for and then playing a role in moving toward them. *We don't wait* until we are sure of success. We don't limit our choices to the outcomes that seem likely. Instead, we focus on what we truly, deeply long for, and then we proceed to take determined steps in that direction. This is the second thread we follow.

We can apply this same principle when looking at ourselves and the different ways we might respond to world issues. We can react to world crises in many ways, with a spectrum of possible responses, from our best to our worst. We can rise to the occasion with wisdom, courage, and care, or we can shrink from the challenge, blot it out, or look away. When we engage Active Hope, we consciously choose to draw out our best responses, so that we might surprise even ourselves by what we bring forth. Can we train ourselves to become more courageous, inspired, and connected? This takes us to the next thread.

The Thread of the Spiral of the Work That Reconnects

The spiral of the Work That Reconnects maps out an empowerment process that journeys through four successive movements, or stations: *coming from gratitude*, *honoring our pain for the world*, *seeing with new eyes*, and *going forth* (see figure 7). This spiral is a source we can come back to again and again for strength and fresh insights. It reminds us that we are larger, stronger, deeper, and more creative than we've grown accustomed to believing. When we come from gratitude, we become more present to the wonder of being alive in this amazing living world, to the many gifts we receive, to the beauty and mystery it offers. Yet the very act of looking at what we love and value in our world brings with it an awareness of the vast violation underway, the despoliation and unraveling.

Coming from gratitude grounds us in trust and psychological buoyancy, which supports us in facing harsh realities in the second station. From gratitude we naturally flow to honoring our pain for the world. Dedicating time and attention to honoring this pain opens up space to hear our sorrow, fear, outrage, and other felt responses to what is happening to our world. Admitting the depths of our anguish, even to ourselves, takes us into culturally forbidden territory. From an early age we've been told to pull ourselves together, to cheer up or shut up. By honoring our pain for the world, we break through the taboos that silence our distress. When the activating siren of inner alarm is no longer muffled or shut out, something gets switched on inside us. It is our survival response.

The term *honoring* implies attentive respect and recognition of value. Our pain for the world not only alerts us to danger but also reveals our profound caring. And this caring derives from our interconnectedness with all life. We need not fear it.

In the third stage, we step further into the perceptual shift that recognizes our pain for the world arises from our love for life. Seeing with new eyes reveals the wider web of resources available to us through our rootedness within a deeper, wider ecological self. This third station draws on insights from holistic science and indigenous wisdom, as well as from our creative imaginations. It opens us to a new view of what is possible and a new grasp of our power to act.

The final station, going forth, involves clarifying our vision of how we can act for the healing of our world and identifying practical steps that move our vision forward.

The spiral offers a transformational journey that deepens our capacity to act for the sake of life on Earth. We call it a spiral rather than a cycle because every time we move through the four stations,

Figure 7. The spiral of the Work That Reconnects as painted by Dori Midnight.

we experience them differently or more fully. Each station reconnects us with our world and can surprise us with hidden gems. As each naturally unfolds into the next, fresh insights and revelations create a flow and build up a momentum. We are letting our world act on us and through us.

The Work That Reconnects as a Personal Practice

The spiral provides a structure we can fall back on, and into, whenever we need to tap into the resilience and resourcefulness arising from the larger web of life. If you're feeling sickened by a disturbing news report, you can step into gratitude simply by focusing on your breath and taking a moment to give thanks for whatever may be sustaining you in that moment. As you feel the air entering your nostrils, give thanks for oxygen, for your lungs, for all that brings you to life. The question "To whom am I grateful?" moves your attention beyond yourself to those who support you.

A moment of gratitude grounds us, strengthening our capacity to look at, rather than turn away from, disturbing information. As you allow yourself to take in whatever you see, allow yourself also to feel whatever you feel. When you experience pain for something beyond your immediate self-interest, this reveals your connection, caring, compassion — your interbeing with all life. By honoring your pain for the world, in whatever form, you take it seriously and allow it to rouse you.

When seeing with new eyes, you know that it isn't just you facing this challenge. You are just one part of a much larger story, a continuing stream of life on Earth that has flowed for more than three and a half billion years and that has survived five mass extinctions. When you sink into this deeper, stronger flow and experience yourself as part of it, a different set of possibilities emerges. Widening your vision brings your awareness to resources available to you, since through the very channels of connectedness that pain for the world flows, so too come strength, courage, renewed determination, and the help of allies.

With the shift of perception that seeing with new eyes brings, you can let go of the need to plan every step; instead, trust your intention and the emerging sequence of events that springs from that. Focus on finding and playing your part, offering your own contribution, your unique gift of Active Hope. As you move into going forth, you consider what this next step will be. Then you take that step.

What we've described here is a short form of the spiral that might take only a few minutes to complete. Like a fractal that has the same characteristic shape at whatever scale it is viewed, the form of the spiral can be applied to a wide range of time frames, with rotations happening over minutes, hours, days, or weeks. We move through the four stations in a way that supports our intention to act for the sake of life on Earth. The more familiar you become with this strengthening journey, the more you will trust it. Each of these stations contains hidden depths, rich meaning, and treasures to explore. It is to these that we turn in the chapters ahead.

Try This: Seven Sentence Starters in Support of Active Hope

Here is a way of journeying round the spiral using a series of seven sentence starters (see figure 8). You can do this as a personal practice — seeing what words come naturally when prompted by each starter, perhaps writing them in a notebook or speaking them to yourself. This practice also works well as a structured conversation in pairs. You might ask a friend, "Hey, would you like to do a spiral together?," then take turns completing the following sentences. The starting point is to look out at what's happening in the world, then see what comes with each of these prompts.

1. I love...
2. I'd like to thank...
3. Looking at the future we're heading into, my concerns include...
4. Facing these concerns, what inspires me is...

5. Looking at the future we're heading into, what I deeply hope for is...
6. A part I'd like to play in support of this is...
7. A step I'll take toward this in the next week is...

When doing this exercise in pairs, we've found that setting a timer for just a minute or two for each sentence adds to the sense of momentum in moving round the spiral. With a minute spent on each part, the conversation takes about half an hour. It also works well to group the last three sentences together, with one partner completing sentences 5, 6, and 7 and then swapping. We have a video demonstrating this practice in pairs at activehope.info/videos.

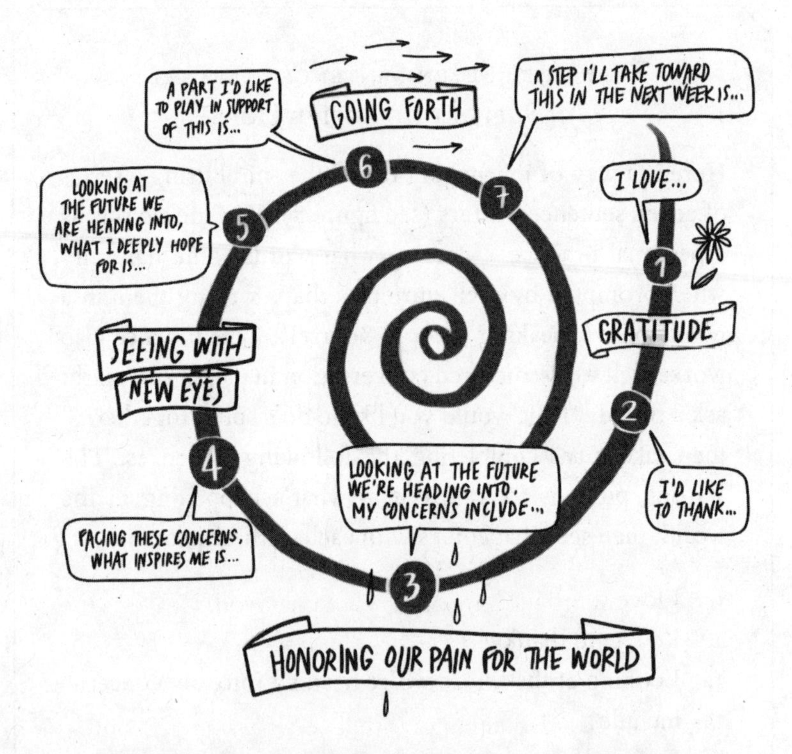

Figure 8. Seven sentence starters in support of Active Hope.

CHAPTER THREE

Coming from Gratitude

"A felt sense of wonder, thankfulness, and appreciation for life" is how the renowned psychologist Robert Emmons defines gratitude.[1] If you're feeling low, it might seem a stretch to focus on something so positive. Yet recognizing the gifts in your life is profoundly steadying, increasing your psychological buoyancy, which helps you maintain your balance and poise when entering rough waters. Put simply, our gladness to be here, when the continuation of life itself is in doubt, kindles our resilience. Our essential wonder at the root of life empowers us to face the unprecedented perils of our time.

Gratitude Promotes a Sense of Well-Being

Recent research has shown that people experiencing high levels of gratitude tend to be happier and more satisfied with their lives.[2] Are they grateful *because* they're happy, or does gratitude *make* people happier? To find out, volunteers were asked to keep gratitude journals in which, at regular intervals, they recorded events they felt thankful for. Controlled trials have shown this simple intervention to have a profound and reliable impact on mood.[3] Indeed, the results are so striking that if a medication were invented with similar benefits, we'd probably see it described as a new wonder drug.

The process of keeping a gratitude journal focuses your attention on ways you're appreciating and being resourced by what's around you and within you. If each evening, before you go to bed, you ask yourself, "What happened today that I'm glad about?," you will start searching for recent memories that bring a glow of contentment. They might be of small things, such as a conversation with a friend, a minute of watching a bird in flight, or the satisfaction of completing a task. When we're busy, moments like these can too easily pass us by, but a gratitude diary builds them into a pool of memories we can be nourished by. Such a simple step trains our minds in a basic capacity for awe and reverence, this countering the undermining impact of panic or paralysis. Experiencing gratitude is a learnable skill that improves with practice. It isn't dependent on things going well or on receiving favors from others. It's about getting better at spotting what's already there.

Try This: A Gratitude Practice

Notice: Scan your recent memories and identify something in the past twenty-four hours that makes you think, "I'm glad that happened."

Savor: Close your eyes and imagine that you are experiencing that moment again. Notice colors, tastes, sounds, smells, and the sensations in your body. Notice also how you feel in yourself.

Give thanks: Who or what helped this moment to happen? If another being was involved, consider how thanks can be offered.

Gratitude has three elements. The first is *appreciation*, the valuing of what has happened. The second is *attribution*, where another's role is recognized. The third is *giving thanks*, where rather than just experiencing gratitude, you give it legs by acting on it.

Gratitude Builds Trust and Generosity

Gratitude nourishes trust, helping us acknowledge the times we've been able to count on one another. By making us more likely to return favors and help others, it also encourages us to act in ways that strengthen the networks of support around us. As psychologists Emily Polak and Michael McCullough point out, "Gratitude alerts us that there are people out there with our well-being in mind, and it motivates us to deepen our own reservoirs of social capital through reciprocation."[4] In such ways it plays a key role in the evolution of cooperative behavior and societies.

Gratitude as an Antidote to Consumerism

While gratitude leads to increased happiness and life satisfaction, materialism — placing a higher value on material possessions than on meaningful relationships — has the opposite effect. In reviewing the research, Polak and McCullough conclude: "The pursuit of wealth and possessions as an end unto itself is associated with lower levels of well-being, lower life satisfaction and happiness, more symptoms of depression and anxiety, more physical problems such as headaches, and a variety of mental disorders."[5]

Affluenza is a term used to describe the emotional distress that arises from a preoccupation with possessions and appearance. Psychologist Oliver James views it as a form of psychological virus that infects our thinking and is transmitted by television, glossy magazines, and advertisements.[6] The toxic belief at the core of this condition is that happiness is based on how we look and what we have.

When women were asked to rate their self-esteem and satisfaction with their appearance, measures for both fell after the women looked at photos of models in women's magazines.[7] How we feel depends so much on what we compare ourselves to, and an increase in eating disorders is one of the consequences of having thin models as a reference group. In 1995, the year television was introduced to Fiji, there were no recorded cases of bulimia on the island. Yet within three years, 11 percent of young Fijian women were found to be suffering from this eating disorder.[8]

Gratitude is about acknowledging the gifts in what you're already experiencing. The advertising industry undermines this by convincing you that you're missing something. On a website for marketing professionals, the advertiser's Law of Dissatisfaction is described like this:

> The job of advertisers is to create dissatisfaction in its audience. If people are happy with how they look, they are not going to buy cosmetics or diet books.... If people are happy with who they are, where they are in life, and what they got, they just aren't customer potential — that is, unless you make them unhappy....
>
> The audience, after seeing what they could look like, is no longer happy with what they do look like, and they are now motivated to buy into the promise of change.[9]

Research shows that people living in countries that spend more on advertising tend to be less satisfied with their lives.[10] Depression has reached epidemic proportions, with one in two people in the Western world likely to suffer a significant episode at some point in their lives.[11] The consumer lifestyle isn't just wrecking our world; it is also making us miserable. Can gratitude play a role in our rehabilitation?

In looking at what drives materialism, psychology professor and researcher Tim Kasser identifies two main factors: feelings of insecurity and exposure to social models expressing materialistic values.[12] Gratitude, by promoting feelings of satisfaction with what you have, counters feelings of insecurity and pulls us out of the rat race. It shifts our focus from what's missing to what's there. If we were to design a cultural therapy that protected us from depression and, at the same time, helped reduce consumerism, it would surely include cultivating our ability to experience gratitude. Training ourselves in the skill of gratitude is part of the Great Turning.

Try This: Open Sentences on Gratitude

Each of these open sentences can act as an invitation to experience gratitude. You can think these sentences to yourself, put them in writing, or try the exercise with a partner, taking turns to speak and listen. It is worth devoting a few minutes or more to each sentence. Whenever you're not sure what to say, come back to the beginning of the sentence and see what naturally follows — it may be different each time you do it.

- Some things I love about being alive on Earth are …
- A place that was magical to me as a child was …
- Some things I love doing are …
- Some things I love making are …
- A person who helped me believe in myself is or was …
- Some things I appreciate about myself are …

Blocks to Gratitude

At times gratitude comes easily. If you're falling in love, having a run of good luck, or just generally delighted with how things are going, appreciation and thankfulness might feel natural. But what if there isn't so much to feel happy about? What about the times when relationships go sour, you experience injury or violation, or the landscape of your life looks bleak?

If you're facing a tragedy in your life or in the world, searching for reasons to be grateful might initially feel uncomfortably close to denial. But you don't have to feel thankful for everything that's happened. It is more a case of recognizing there's always a larger picture, a bigger view, and that it contains both positive and negative aspects. To find our power to see the hard parts clearly and respond constructively, we need to draw on resources that bring out the best in us. Gratitude does this. It's a resource we can learn to tap into at any moment.

Here's an example. Julia had just seen the news. She felt outraged. A school in a refugee camp had been bombed, children were killed, and she was so full of fury she could hardly speak about anything else. Then, for a moment, she thought of the reporters who had covered this story. They had risked their lives so that she could be kept informed. The news editors may also have stuck their necks out, choosing to include this item in their program rather than the latest celebrity gossip. As she thought about the steps they had taken, she felt grateful. Her gratitude reminded her she wasn't alone in caring about what happened.

When violations and injustices occur, trust is often a casualty. Loss of trust makes it harder to experience gratitude; even when help is given, the distrustful part of us may wonder what agendas are hidden behind the support. When trust is broken or undermined, is it possible to rebuild it? Trust and gratitude feed each other: to deepen our capacity for thankfulness in difficult times, we need to learn from those who have mastered this quality.

LEARNING FROM THE HAUDENOSAUNEE

In autumn 1977, delegates from the Haudenosaunee, Native Americans also known as the Iroquois Confederacy, traveled to a UN conference in Geneva, Switzerland. They had a warning and a prophecy to share, presenting it alongside a description of their core values and view of the world. Their "Basic Call to Consciousness," as it is known, contained the following paragraph:

> The original instructions direct that we who walk about on the Earth are to express a great respect, an affection, and a gratitude toward all the spirits which create and support Life. We give a greeting and thanksgiving to the many supporters of our own lives — the corn, beans, squash, the winds, the sun. When people cease to respect and express gratitude for these many things, then all life will be destroyed, and human life on this planet will come to an end.[13]

The Haudenosaunee regard gratitude as essential in a world they understand as sustained by the interdependence of all things. From the perspective of Western individualism, that view might seem hard to grasp. In the Business as Usual story, self-made success is among the most highly prized of victories. If we can stand on our own two feet, why give thanks to the beans and the corn? Yet the notion that we can be completely independent or self-made denies the reality of our interdependence with other people and our natural world.

The Haudenosaunee see humans as interwoven strands of a far larger web of life. Plants, trees, rivers, and the sun are respected and thanked as fellow beings in a larger community of mutual aid. If you have this view of life, you don't tear down the forests or pollute the rivers. Instead, as their "Basic Call to Consciousness" describes, you accept other life-forms as part of your extended family. "We are shown that our life exists with the tree life, that our well-being depends on the well-being of the Vegetable Life, that we are close relatives of the four-legged beings."[14]

The Haudenosaunee's thanksgiving is known as "the words

that come before all else" and precedes every council meeting. Instead of being reserved for a special day each year, thanksgiving becomes an essential element in their way of life. What is striking about their thanksgiving practices is that the Haudenosaunee don't focus on possessions or personal good fortune, but on blessings we all receive from the natural world. While the expressions of gratitude are delivered spontaneously in people's own words, the order and format follow a traditional structure. A version of the thanksgiving address has been published by the Mohawk, one of the six nations of the Haudenosaunee, and begins like this:

The People

> Today we have gathered and we see that the cycles of life continue. We have been given the duty to live in balance and harmony with each other and all living things. So now, we bring our minds together as one as we give greetings and thanks to each other as people.
>
> Now our minds are one.

The Earth Mother

> We are all thankful to our Mother, the Earth, for she gives us all that we need for life. She supports our feet as we walk about upon her. It gives us joy that she continues to care for us as she has from the beginning of time. To our mother, we send greetings and thanks.
>
> Now our minds are one.[15]

As the words of thanks continue, they are directed to the waters of the world; the fish in the water; plant life, including food plants and medicine herbs; the animals; the trees; the birds, who each day "remind us to enjoy and appreciate life"; the winds; the thunder beings and lightning, who "bring with them the water that renews life"; our eldest brother, the Sun; our Grandmother Moon, who "governs the movement of the ocean tides"; the stars "spread across the sky like jewelry"; enlightened Teachers; and the Creator, or Great Spirit. Finally, "if something was forgotten, we leave it to each individual to send such greetings and thanks in their own way." After every element of our thanking is mentioned, the words are reverently repeated, "And now our minds are one."[16]

Thanksgivings like this one deepen our instinctual knowledge that we belong to a larger web and have an essential role to play in its well-being. As Haudenosaunee Chief Leon Shenandoah said in his address to the General Assembly of the United Nations in 1985, "Every human being has a sacred duty to protect the welfare of our Mother Earth from whom all life comes."[17]

Different stories give us different purposes. In the Business as Usual model, nearly everything is privatized. The parts of our world remaining outside individual or corporate ownership, such as the air or the oceans, are not seen as our responsibility. Gratitude is viewed as politeness, not necessity. In their "Basic Call to Consciousness," the Haudenosaunee tell a very different story, one in which our well-being depends on the natural world and gratitude keeps us close to the purpose of taking care of life. When we forget this, the larger ecology we are embedded in recedes from sight — and the world unravels.

The Modern Science of Gaia Theory

Just as we depend on plants for food, we also rely on them to make the air breathable. Our two neighboring planets, Mars and Venus, have atmospheres that would kill us in minutes, and we've only recently discovered that Earth's atmosphere used to be the same. Three billion years ago, our planet's air, like that of Mars and Venus, had vastly more carbon dioxide and hardly any oxygen.[18] Over the next two billion years or so, early plant life did us the remarkable service of making our atmosphere breathable by generating an abundance of oxygen and removing much of the carbon dioxide.

Oxygen is a highly reactive gas, which wouldn't normally be expected to exist at levels as high as the 20 percent we have now. It was the chemically unlikely fact that oxygen has remained at this level for hundreds of millions of years that led British scientist James Lovelock to develop the hypothesis that would become Gaia theory: that our Earth is a living, self-regulating system. Here is how he described his moment of insight:

> An awesome thought came to me. The Earth's atmosphere was an extraordinary and unstable mixture of gases, yet I knew that it was constant in composition over quite long periods of time. Could it be that life on Earth not only made the atmosphere, but also regulated it — keeping it at a constant composition, and at a level favourable for organisms?[19]

The core tenet of Gaia theory, that our planet is a self-regulating system, parallels the way our bodies keep arterial oxygen and temperature levels stable or the way termite colonies maintain their internal temperature and humidity. This capacity of living systems to keep themselves in balance is known as *homeostasis*. Gaia theory shows how life looks after itself, different species acting together to maintain the dynamic balance of nature.

As stars grow older, they tend to burn brighter. Because of this, it is estimated that our sun now puts out at least 25 percent more heat than it did when life began on Earth 3.5 billion years ago.[20] Yet has our planet become 25 percent hotter? Human life wouldn't exist if it had. And we have plant life to thank for this. By absorbing carbon dioxide, plants reduce the greenhouse effect of this gas, keeping the planetary temperature within a range suitable for complex life such as ours.

Giving Back and Giving Forward

A timber executive once remarked that when he looked at a tree, all he saw was a pile of money on a stump. Compare this with the Haudenosaunee view that trees should be treated with gratitude and respect. If we understand how essential trees are to our very breathing, we want to support them. Such a dynamic pulls us into a cycle of regeneration, in which we take what we need to live and also give back. Because our industrialized culture has forgotten this principle of reciprocity, forests continue to shrink and deserts to grow. To counter the unraveling, we must develop an ecological intelligence that recognizes it takes two to tango. It's a dance to the tune of gratitude.

Try This: Thanking What Supports You to Live

Next time you see a tree or a plant, take a moment to express thanks. With each breath you take in, experience gratitude for the oxygen that would simply not be there save for the magnificent work plants have done in transforming our atmosphere and making it breathable. As you look at all the greenery, bear in mind also that plants, by absorbing carbon dioxide and reducing the greenhouse effect, have saved our world from becoming dangerously overheated. Without plants and all they do for us, we would not be alive today. Feel the physical touch and rhythm of this gratitude as your lungs fill and exhale.

The carbon dioxide released when we burn fossil fuels returns to our living atmosphere the gas that ancient plants removed hundreds of millions of years ago. Our excessive use of fossil fuels reverses a cooling mechanism of our planet, and temperatures are rising. Today's forests help planetary cooling not only by absorbing carbon dioxide but also by releasing a substance that helps clouds form.[21] Clouds help reduce overheating and add to local rainfall. When tropical forests are logged and burned, the local climate becomes hotter and drier, making it harder for trees to grow again. Tropical forests like the Amazon are under assault not only from deforestation but also from drought related to climate change. They need our help, just as we need theirs.

Much of the oxygen we breathe comes from plants that died long ago. We can give thanks to these ancestors of our present-day foliage, but we can't give back to them. When we are unable to return a favor, however, we can pay it forward to someone or something else.[22] Using this approach, we can see ourselves as part of a larger flow of giving and receiving through time. Receiving from the past, we can give to the future. When we are tackling issues such as climate change, guilt and fear need not be our only motivation. Gratitude can be equally powerful and fundamental to the self-organizing intelligence that supports life.

CHAPTER FOUR

Honoring Our Pain for the World

Parsifal, a popular knight in the court of King Arthur, was out on a quest and found himself riding for days through a strange, devastated region, a wasteland where nothing grew. Searching for food and shelter, he came to a lake, where he saw two figures fishing from a boat, one of them dressed in royal garments.

"Where can I find lodging?" Parsifal called out.

The regally dressed fellow, who was the Fisher King, invited him to stay at his castle nearby and gave directions. Before long, Parsifal found himself at the great castle of the Fisher King. Welcomed by courtiers, Parsifal was escorted to a hall where a ceremonial banquet was underway. There, on a kind of hammock bed, lay the king, who appeared to be in great pain from a wound in his groin.

The king was victim of a spell that had stopped his wound from healing. This spell had taken from the ruler and his lands the powers of regeneration. Yet as the evening wore on, no one in the court made mention of the horror they faced.

It was known to them, however, that the spell might be broken if and when a knight appeared and asked the king a compassionate question. But Parsifal had been brought up to believe that it was rude, even out of order, to ask questions of people in authority. So even though he approached the king to greet him and could smell the foul stench of his wound, he acted as if nothing was wrong.

The next morning, Parsifal woke up to find that the king, his courtiers, and the entire castle had disappeared. Setting off, he eventually found his way out of the wasteland and made it back to King Arthur's court. A feast was held to mark the return of this much-admired knight. The celebration was interrupted by the dramatic entrance of a hideous but wise witch called Cundrie. She publicly scolded Parsifal for lacking the compassion to ask about the suffering of the Fisher King and his people.

"You call yourself a knight," Cundrie cried out. "Look at you! You didn't even have the guts to ask what was going on."

Haunted by Cundrie's accusations and his own feelings of failure, Parsifal fell into a depression. He finally realized he had to return to the court of the Fisher King. It took many years, and many adventures, to find the mysterious castle again. When he did, he found the king in a state of even greater suffering.

This time Parsifal strode right up to the king and knelt before him. "My lord, what aileth thee?" he asked. That Parsifal cared enough to ask this question had an immediate and powerful effect. Color returned to the cheeks of the king, who rose from his couch, his health regained. In the same moment, the wasteland began once again to bloom. The spell was broken.

Parsifal's story, which is part of the legend of the Holy Grail, was put into writing more than eight hundred years ago.[1] Yet the situation it describes, one of great suffering not being acknowledged and of people carrying on as if nothing were wrong, is easily recognizable today. The question "What aileth thee?" — or its modern equivalent, "What troubles you?" — invites people to express their concern and anguish. There are all kinds of reasons why this question doesn't get asked and why, like Parsifal, we sometimes find ourselves caught up in the pretense that all is well, even when we know it isn't. When this happens in relation to the state of our world, as it so often does, our survival response becomes blocked.

This chapter explores what gets in the way of acknowledging disturbing realities and looks at how honoring our pain for the world can break the spell that maintains Business as Usual.

Understanding the Blocked Response

A key survival mechanism in our response to danger is the activation of alarm. A fire chief once remarked that when a building burns, the people most at risk are not those who panic, who tend to get out quickly, but those who, not having grasped the degree of emergency, sort through their possessions deciding which to save. With this in mind, psychologists Bibb Latané and John Darley carried out a study examining factors that influence our response to potential danger.[2]

In the study, volunteer subjects were asked to wait in a room and fill out a questionnaire. While they were following these instructions, a steady stream of smokelike vapor started pouring in through a vent in the wall and filling the room. If individuals were in the room alone, they responded quickly, leaving the room and looking for help. But when three people sat in the room together, they looked to see how others responded. If they saw others remaining calm and continuing to fill in their questionnaires, they were more likely to do so too. Even when the room became so smoky that it was difficult to see and some subjects were coughing or rubbing their eyes, nearly two-thirds of the subjects persisted with their questionnaires. This tendency to quietly conform was even more pronounced if the group of three contained two actors instructed to ignore the smoke. Under these conditions, 90 percent of subjects continued filling in their forms for a full six minutes before they were "rescued" by one of the researchers.

This experiment serves as a metaphor for our responses to our present planetary emergency. If the smoke represents disturbing information, filling out the questionnaire is like continuing with

Business as Usual. If we are to survive as a species, we need to understand how our active responses to danger get blocked and also how we can prevent this from happening. What gets in our way of noticing the smoke? Often we experience an inner tension between the impulse to do something and the resistance to it. There are many varieties of resistance. Here are seven common ones.

1. I don't believe it's that dangerous.

When the researchers interviewed the people who did not respond to the smoke, a common reason they gave was that they didn't believe there was a fire. They had developed another explanation for the smoke that led them to discount it as a cause for alarm. That this happened so much more when the subjects saw others ignoring the smoke demonstrates how powerfully influenced we are by what we observe others doing. We notice this same dynamic at play with global events as well: if we learn some disturbing information about an issue but see most people acting as if it is no big deal, then it is easier to believe the problem can't be that serious.

Corporate-controlled media amplifies this effect. How often, for example, have you seen characters in TV soaps, sitcoms, or dramas express concern about the condition of our world? Most major television stations and newspapers are owned by a tiny number of corporations, and their main source of revenue is advertising. Their commercial goal is to present audiences to advertisers. As a result, crucial information about the unraveling of our world is downplayed or entirely omitted in news reports. When, for example, the World Meteorological Organization announced in December 2009 that the decade 2000–2009 was on track to be the warmest on record, the managing editor of Fox News sent the following email directive, telling news teams and producers to "refrain from asserting that the planet has warmed (or cooled) in any given period without IMMEDIATELY pointing out that such theories are based upon data that critics have called into question. It is not our place as journalists to assert such notions as facts."[3]

2. It isn't my role to sort this out.

The individualism of Business as Usual draws a dividing line between our problems and those of other people and an even sharper line between the concerns of humans and those of other species. "What happens to polar bears is really not my responsibility," said Martin, a neighbor, when discussing climate change. Such fragmentation of responsibility, with its diminishing care for our world, is the dominant mode in industrialized societies. In this view, the world is divided into separate pieces, and we're responsible only for the pieces we own, control, or inhabit. We may take great care of our backyards but view anything beyond the end of the street as someone else's problem.

3. I don't want to stand out from the crowd.

When Joanna first heard the shocking news that President John F. Kennedy had been assassinated, she was standing in the checkout line at a supermarket. Although the radio announcement had been clearly audible, people carried on shopping as if nothing had happened. For reasons she still cannot fathom, Joanna, like those around her, remained passively in line without speaking.

The invisible pressure to conform is both subtle and powerful. It can even lead us to doubt our perceptions, for if no one else is acknowledging a problem, perhaps we've got it wrong. By naming an issue, on the other hand, we bring it into the open. In doing so we disturb the status quo, since it is more difficult to ignore a problem once it has been acknowledged. Whistleblowers are often blamed for the inconvenience of our being forced to address previously ignored problems. When speaking out is seen as too risky, a climate of fear can lead to a culture of collusion. Tony, a participant in one of our courses, described his experience of working in a large corporation: "In my work, it is common practice to leave your conscience at the door. Questioning company policy is viewed as disloyalty; it could threaten your career." Especially at a time of job insecurity, the fear of being labeled a troublemaker discourages the expression of disturbing information and alarm.

4. This information threatens my commercial or political interests.

Throughout the 1990s, Thailand's chief meteorologist, Smith Thammasaroj, gave repeated warnings about the risk of a tsunami to the country's southwest coast.[4] For decades he had studied the relationship between underwater earthquakes and these giant tidal waves; he was so concerned that he recommended beachside hotels in the coastal provinces of Phuket, Phang Nga, and Krabi be fitted with tsunami alarms. His cautions were seen as harmful to the tourist trade, he was sidelined from his job, and his recommendations were ignored. On December 26, 2004, an underwater earthquake off Sumatra led to a massive tsunami striking the very regions he had warned about. In one of the worst disasters in recorded history, more than 5,000 died in Thailand and more than 225,000 in thirteen other countries.

Before the tsunami struck, Thammasaroj, accused of scaremongering, was even banned from some of the provinces he had been trying to protect. One way we resist information we don't want to hear is to attack its source. The tobacco industry did this for decades, with well-funded campaigns to discredit both the science and the scientists proving that smoking killed people. Even though internal memos show the industry knew the science was correct, they still sought to invalidate research findings that clashed with their commercial interests.[5] Public relations companies were paid millions to cast doubt on the link between tobacco and health problems. The same thing is happening today with the science of climate change: oil and gas companies spent more than $1 billion on lobbying and misinformation to block climate action in the three years after the Paris Agreement of 2015.[6]

5. It is so upsetting that I prefer not to think about it.

While human history has seen its share of disasters, what's different these days is that well-established trends set us on a course toward catastrophe. As our world continues to heat up, the potential outcome is horrifying. So why is the dominant response to continue with Business as Usual? One reason is that our planetary emergency can feel too painful even to contemplate. Here are some of the comments made by people we interviewed. From Isabelle, the mother of two young children: "I used to be politically active, but these days I can't bear to listen to the news because I find it so depressing." And from John, a grandparent: "I know the world's a mess, but I don't want to be constantly reminded of it. I prefer not to think about it."

6. I feel paralyzed. I'm aware of the danger, but I don't know what to do.

How can we begin to tackle our problems if we find them too painful even to think about? Yet being aware of the problems and not knowing how to respond can be even more agonizing. Here is how Jennifer, who'd been an activist for many years, described her experience: "Years ago, I'd felt optimistic; I sensed people were waking up, that there was a growing determination to do something. But it didn't come to much, and the problems we face just seem to get worse. I feel I must do something; I can't just watch this. Yet there is so much that needs doing, I don't know where to start. It is overwhelming, and I feel paralyzed."

7. There's no point doing anything, since it won't make any difference.

What if we're too late? What if we've already done so much damage that our industrialized society, like the *Titanic*, is sinking and our world is dying? This perspective makes it hard to find authentic meaning and purpose in life. On the one hand, we recognize that the Business as Usual mode is based on living a lie. Yet on the other, we view attempts to bring positive change with cynicism and see those who engage in them as peddling false optimism. What remains is a story of decline, of going through the motions of living, of making the best of what we have before or during the final collapse.

Though this outlook seems bleak, an international survey of more than ten thousand young people in 2021 suggests such feelings are widespread. Over half (56 percent) said they thought humanity was doomed, and four out of ten were hesitant to have children because of their concerns.[7]

WORKING WITH OUR DESPAIR FOR THE WORLD

The first four resistances described here grow out of the Business as Usual story that doesn't recognize the crisis; the last three are linked to the Great Unraveling and the ghastly realization that things are so much worse than we'd thought.

We can exist in both these realities at the same time — going about our normal lives in the mode of Business as Usual while also remaining painfully aware of the multifaceted crisis unfolding around us. Living in this double reality creates a split in our mind. It is as though one part of our brain operates on the default assumption that things are fine, while another part knows full well they are not. One way of dealing with the confusion and agony of this split is to push the crisis out of view, like the people in the study who carried on with their questionnaires while trying not to think about the smoke filling the room. But this way of living is difficult to sustain, particularly as the condition of our world continues to worsen.

It is hard even to talk about this. Taboos in normal conversation block the discussion of anything considered too depressing. When we feel dread about what may lie ahead, outrage at what is happening to our planet, or sadness about what has already been lost, it is likely that we have nowhere to take these feelings. As a result, we tend to keep them to ourselves and suffer in isolation. If people see us keeping quiet, not mentioning the crisis, then, as in the smoke-filled room, they'll be more likely to keep silent as well. Yet if we reveal our distress and name the horrors we see, we're likely to be called overly negative or too emotional.

We can be caught between two fears — the fear of what will happen if we, as a society, continue the way we're going and the fear of acknowledging how bad things are because doing so brings up such despair. If we listen to the first fear, it can provoke responses that aid our survival; but to benefit from the wake-up call, we need to free ourselves from the stifling effect of the second fear. There are ways to do this.

More than forty years ago, Joanna developed an approach that helps people respond creatively to world crises rather than feeling overwhelmed or paralyzed by distress. As with grief work, facing our distress doesn't make it disappear. Instead, when we do face it, we are able to place our distress within a larger landscape that gives it a different meaning. Rather than feeling afraid of our pain for the world, we learn to feel strengthened by it.

When she first developed it in the late 1970s, this approach, which largely involves working with groups in a structured workshop format, was known as "despair and empowerment work." Because it deepens our relationship with the web of life, the term "deep ecology workshops" has also been used. Since the late 1990s, this approach has been known as the Work That Reconnects.[8] The principles and practices can be applied not just to workshops but also to education, psychotherapy, community organizing, one-to-one conversations, personal development, and spiritual practice. At the heart of the Work That Reconnects lies a different way of thinking about our pain for the world.

Pain for the World Is Normal, Healthy, and Widespread

Many years ago, a therapist told Joanna that her strong feelings in response to a demolished forest stemmed from her fear of her libido, as represented by the bulldozers. This tendency to reduce our distress about planetary conditions to some psychological problem or neurosis is common. Our anguish and alarm about what we're doing to our world are viewed as symptoms to be treated or as markers of an underlying personal issue. Self-help groups have not been immune to this approach: a recovering alcoholic, mentioning his fears for our world in his support group, found himself challenged with the question "What are you running away from in your own life that you bring up such concerns?"

Drawing on her doctoral studies exploring convergences between Buddhist philosophy and systems theory, Joanna saw that the meaning we give to our emotional responses is of central importance. The perception of radical interconnectedness found in both Buddhism and systems thinking supports a reframing of our distress about world conditions. It helps us recognize how healthy a reaction this distress is and how necessary it is for our survival. A central principle of the Work That Reconnects is that *pain for the world* — a phrase that covers a range of feelings, including outrage, alarm, grief, guilt, dread, and despair — is a normal, healthy response to a world in trauma.

In Buddhism, as in other major world religions, open alertness that allows our heart to be stirred by the suffering of others is appreciated as a strength. Indeed, in every spiritual tradition, compassion, which literally means "to suffer with," is prized as an essential and noble capacity. It is evidence of our interconnectedness with all life.

The concept of negative feedback loops from systems theory helps us recognize how this ability to suffer with our world is essential for our survival. We navigate through life by paying attention to information, or feedback, that tells us when we are off course and by responding with a course correction. This dynamic process loops continually: we stray off course, notice that, make a response that brings us back on course, stray off course again, notice again, come back on course again (see figure 9). Since the process functions to diminish the degree to which one is off course, it is called a *negative feedback loop*. It is also sometimes referred to as a *balancing feedback loop*.

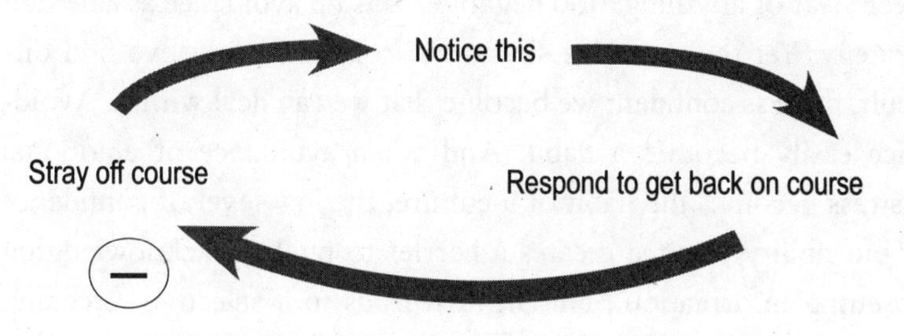

Figure 9. A negative feedback loop:
each response reduces deviation from the chosen course.

It is through loops like this that living systems keep themselves in balance. For example, if we feel too cold, we might respond by putting on a jacket. If, after a while, we feel too hot, we take the jacket off. If a roomful of people start to feel too hot, someone might open a window. In this case, the system of the group acts through its members to keep the room temperature comfortable.

How would we notice if we were straying off course as a society? We would start to feel uncomfortable. If we were heading in a dangerous direction, we might feel alarmed; if something unacceptable was taking place, we might feel outraged. If parts of our world that we loved were dying, we would expect to grieve. These feelings are normal, healthy responses. They help us notice what's going on; they are also what rouses us to respond.

Unblocking Feedback Releases Energy

In our workshops we see people sharing their anguish and despair for our world situation and then emerging energized and nourished by the experience. Facing and expressing painful feelings enlivens people, often releasing strong team spirit and even hilarity.

A strong current in mainstream culture views depressing news, gloomy thoughts, and feelings of distress as "negative experiences" from which we need to protect ourselves. The notion that we should steer clear of anything "too negative" sets up avoidance as a default strategy. Yet the more we shy away from something we find difficult, the less confident we become that we can deal with it. Avoidance easily becomes a habit. And when avoidance of emotional distress becomes the habit of a culture, the low level of confidence in our ability to cope creates a barrier to publicly acknowledging upsetting information. This in turn leads to a selective screening out of aspects of reality that seem too painful to bear, too distressing to contemplate.

The disturbing realities of early twenty-first-century Earth, along with what we are likely to face over the coming decades, far exceed any previous experience of disaster in our history. It is not surprising that the feelings provoked by this reality are difficult to bear. "If I open the door to those feelings," said a friend in a recent conversation, "I'll be plunged to a place so low I don't think I'll ever get back." What makes many of us nervous about looking at our grief or despair is the fear that we will become stuck in those emotions. A key issue here is our capacity to deal with distress.

Emotional distress can be motivating, but if it goes beyond what we imagine we can cope with, we may just shut down. While on the outside we may appear to be holding it together, when we close off emotionally we feel less alive, our energy sagging and our sensitivity dulled. We may feel we're just going through the motions. Alcohol, recreational drugs, shopping, and antidepressants are among the devices we use to keep distress under wraps. In the short term, these salves can seem effective. But as we become dependent on them, our society continues to stray off course, and our world becomes a wasteland.

The Work That Reconnects offers a different way forward. Whether in its workshop form or in partnered conversations, it offers a safe, supportive forum where distress can be voiced, heard, and valued, thus unblocking the crucial feedback loop. As one participant said: "I want to face the truth of what's happening in our world. If that truth is painful, then I need to face the pain."

It is our consistent experience that as people open to the flow of their emotional experience, including despair, sadness, guilt, fury, or fear, they feel a weight being lifted from them. In the journey into the pain, something foundational shifts; a turning occurs.

When we touch into our depths, we find that the pit is not bottomless. When people are able to tell the truth about what they feel, see, and know is happening to their world, a transformation occurs. They experience an increased determination to act and a renewed appetite for life.[9]

A range of factors acts together to bring about this shift. First, as noted, repressing information and emotions dampens our energy. It is enlivening to go with, rather than against, the flow of our deep-felt responses to the state of the world. Second, we feel tremendous relief on realizing our solidarity with others. Two other factors play a crucial role. One concerns the process by which we integrate painful information, and the other relates to what pain for the world reveals about the nature of the self.

Information Is Not Enough

Environmental and social change organizations often fall into the trap of assuming that if only people knew how bad things were, they would become active in tackling the issues. So their emphasis falls on presenting information and arguments, backing these up with shocking facts, graphs, and images. This raising of awareness is important, but what if people are already so overwhelmed they are not able to cope with any more distress? Or what if they believe they need to protect themselves from negative thinking? In these cases, presenting more shocking facts can just increase their resistance.

Elisabeth Kübler-Ross, a psychiatrist known for her work with people suffering from life-threatening conditions, described a process of coming to terms with bad news that involves a number of stages.[10] In the beginning, the loss doesn't feel real. Someone may acknowledge that something has happened but act as if this fact hasn't yet been accepted. The information may be taken in on a superficial level but not yet digested.

In his classic text on grief counseling, psychology professor J. William Worden identifies the initial tasks of grieving as first accepting the loss and then feeling the pain of grief.[11] When we feel this emotion, we know not only that the loss is real but also that it matters to us. That's the digestion phase — when the awareness sinks to a deeper place within us so that we take in what it means. Only then can we find a way forward that is based on an accurate perception of reality.

There is a journey to reaching the point of really getting it that our planet faces a life-threatening condition. If we haven't fully taken in and digested the bad news of our planetary predicament, then we might see little reason to step outside the story of Business as Usual. But if the disturbing diagnosis of unraveling has really landed inside us, then we know we can't go back to pretending that all is well.

Each day we lose valuable parts of our biosphere as species become extinct and ecosystems destroyed — yet where is their funeral service? If our world is dying piece by piece without our publicly and collectively expressing our grief, we might easily assume that these losses aren't important. Honoring our pain for the world is a way of valuing our awareness, first, that we have noticed, and second, that we care. Intellectual awareness by itself is not enough. We need to digest the bad news. That is what rouses us to respond.

So what helps us digest the information we're aware of intellectually but haven't yet taken in at a deeper level? A starting point is just to hear ourselves speak what we already know. By giving voice to our concerns, we bring them into the open.

Try This: Open Sentences on Concern

Sit in pairs, with one partner acting as listener and the other as speaker. The role of the listener is to listen attentively without saying anything. For each of the following open sentences, the speaker talks for two minutes or so, saying the first part of the sentence and continuing to express whatever naturally follows. If not sure what to say, the person talking can come back to the beginning of the sentence and start again. Swap roles after each sentence so that both partners get a chance to hear each other, or allow one person to go through all three, then change roles.

- When I see what's happening to the natural world, what breaks my heart is...
- When I see what's happening to human society, what breaks my heart is...
- What I do with these feelings is...

Some unspoken rules govern the content and emotional bandwidth of normal conversations. If we stray too far outside agreeable territory — for example, by raising an issue seen as too disturbing or by expressing a feeling with too much intensity — we are likely to be met by responses that close down our expression. We might find ourselves caught up in an argument or see the issue ignored as conversation is steered back to more comfortable topics. To address the challenges we face, we need to develop ways of talking about them that don't become battles about who is to blame or an avoidance mechanism that filters out the true depths of our concern. By giving each speaker space to talk without interruption, the open sentences process makes room for a different kind of conversation to take place.

This simple exercise often brings up strong feelings. Yet the process is not about *making* people feel something; rather, it is about freeing what is already there. It is making space to hear one another and allow what's under the surface to be expressed. When we allow feelings to move through us, we are less likely to get stuck in them.

When we experience pain for the world, the world is feeling through us. This is a central insight of the work: if feelings flow into us from the world, they can also flow back out again — they don't have to get stuck in us. A process called Breathing Through, adapted from an ancient Buddhist meditation for developing compassion, reinforces this understanding.

TRY THIS: BREATHING THROUGH

Closing your eyes, focus on your breathing. Don't try to breathe in any special way, slow or long. Just watch the breathing as it happens, in and out.

Note any accompanying sensations at the nostrils or in the chest or abdomen. Stay passive and alert, like a cat by a mousehole...

As you watch the breathing, note that it happens by

itself, without your will, without your deciding each time to inhale or exhale. It's as though you're being breathed — being breathed by life. Just as everyone in this room, in this city, on this planet now, is being breathed by life, sustained in a vast living, breathing web...

Now visualize your breath as a stream or ribbon of air. See it flow up through your nose, down through your windpipe, and into your lungs. Now take it through the heart. Picture it flowing through your heart and out through an opening there to reconnect with the larger web of life. Let the breath stream as it passes through you and through your heart, appearing as one loop within that vast web, connecting you with it...

Now open your awareness to the suffering in the world. For now, drop all defenses and open to your knowledge of that suffering. Let it come as concretely as you can... images of your fellow beings in pain and need, in fear and isolation, in prisons, hospitals, tenements, refugee camps... No need to strain for these images; they are present to you by virtue of our interexistence. Relax and just let them surface... the countless hardships of our fellow human beings and of our animal brothers and sisters as well, as they swim the seas and fly the air of this planet...

Now breathe in the pain like invisible tears in the stream of air, up through your nose, down through your trachea, lungs, and heart, and out again into the world... You are asked to do nothing for now but let it pass through your heart... Be sure that stream flows through and out again, don't hang on to that pain... Surrender it for now to the healing capacity of life's vast web.

"Let all sorrows ripen in me," said Shantideva, the Buddhist saint. We help them ripen by passing them through our hearts... making good, rich compost out of all that grief... so we can learn from it, enhancing our larger, collective knowing...

If no images or feelings arise and there is only blankness, gray and numb, breathe that through... The numbness itself is a very real part of our world...

And if what surfaces for you is not pain for other beings so much as losses and hurts in your own life, breathe those through too... Your own difficulties are an integral part of the grief of our world...

Should you fear that with this pain your heart might break, remember that the heart that breaks open can hold the whole universe. Your heart is that large. Trust it. Keep breathing.

A Different View of Self

The Vietnamese Zen master Thich Nhat Hanh was once asked what we need to do to save our world. "What we most need to do," he replied, "is to hear within us the sounds of the Earth crying."[12] The idea of the Earth crying within us, or through us, doesn't make sense if we view ourselves only as separate individuals. Yet if we think of ourselves as deeply embedded in a larger web of life — as Gaia theory, Buddhism, and many other, especially indigenous, spiritual traditions suggest — then the idea of the world feeling through us seems entirely natural.

This view of the self is very different from that found in the extreme individualism of the Business as Usual model. The latter takes each of us as a separate bundle of self-interest, with motivations and emotions that make sense only within the confines of our own stories. Pain for the world tells a different story, one about our interconnectedness. We feel distress when other beings suffer because, at a deep level, we are not separate from them. The isolation that splits us from the living body of our world is an illusion; the pain breaks through to tell us who we really are.

What we see happening in the Work That Reconnects workshops is more than just people learning not to fear their pain for the world, more than just the building of community and the unblocking of feedback. We witness a profound shift in people's relationships with the world, bringing a deep and nourishing sense of being held.

Daniel, a physician, said, "I never felt I belonged anywhere before. Now I know I belong to this world, and in my bones I feel I am part of it." Annie, a schoolteacher, added, "After the workshop, I remember feeling an uncommon and wonderful sense that I do have a place in the world."

Our pain for the world arises out of our interexistence with all life. When we hear the sounds of the Earth crying within us, we're unblocking not just feedback but also the channels of felt connectedness that join us with our world. These channels act like a root system, opening us to a source of strength and resilience as old and enduring as life itself.

In the process, the view that we are separate from our world falls away. The term *deep ecology*, coined by Norwegian ecophilosopher Arne Næss, captures the essence of this shift.[13] When we perceive our deeper identity as an ecological self that includes not just us but also all life on Earth, then acting for the sake of our world doesn't seem like a sacrifice. It seems a natural and fulfilling thing to do.

Personal Practices to Honor Our Pain for the World

Open Sentences and the Parsifal Question

A starting point for exploring your emotional reactions to world conditions is simply to ask yourself how you feel. You may recall that Parsifal broke a spell by asking the king the compassionate question "What aileth thee?" Adapted for our current context, you would ask yourself, "What troubles me about what's happening in our world?," then make room to listen inside for your response. You can draw out your responses more by writing them down. Whenever you're not sure what to say or write, you can use the open sentences technique to give you a starting point and then continue the sentence with whatever naturally flows. The open sentences on page 69 offer a useful launching point for reflection or personal journaling. Some other open sentences we often use in workshops are:

- When I imagine the world we will leave those who come after us, it looks like ...
- One of my worst fears about the future is ...
- The feelings about this that I carry around with me are ...
- Ways I avoid these feelings include ...
- Some ways I can use these feelings are ...

Each open sentence is a springboard into territory usually excluded from conversation. When we bring our fears into the open, they lose their power to haunt us. Jade, a young woman attending one of our workshops, described the impact this had on her:

> Previously I'd been terrified of the prospect of global and ecological collapse. It was so horrifying a thought that I couldn't face it. It was too much to look at world problems, so I tried to shut them out.
>
> At that workshop, things changed. I faced my horrors and remember thinking, "This is my terror, and it is really bad." Previously I'd been really afraid, but afterward my nightmares stopped. I felt more accepting, putting things in perspective better, thinking that we just do our best, so I got on with it.

After the workshop, Jade left her job and started working for an environmental organization. Facing the awfulness had strengthened her resolve to find and play her part in addressing her concerns about the world.

Breathing Through

The Breathing Through meditation described on pages 70–72 helps bring feelings to the surface; it can also become a trusted ally if you fear becoming overwhelmed. Bringing your attention back to your breath has a stabilizing effect, as does remembering the image of feelings coming in, passing through your heart, and then being released into the healing resources of the web of life.

Creative Expression of Our Pain for the World

If we are not used to articulating our emotions, it can sometimes be difficult to know what we feel. A way of drawing out feelings so that they become easier to recognize is to give them a form, such as a shape or color on paper, a sound with your voice, or a texture with clay. Moving beyond words or diving beneath them takes us to that subliminal level where we sense threats, so that we can touch into and release our deeper responses to the condition of our world.

Try This: Drawing Out Your Concerns

Take a blank piece of paper and some colored pens. Scribble, doodle, or draw any images that represent concerns you have and the feelings that accompany them.

When images surface, they allow us to step into the realm of our intuitive responses to our planet time. We may be surprised by what our hands portray: images of concern revealed on paper that we may not have spoken or even been aware of. Images, once birthed, have a reality of their own that we don't need to explain or defend or apologize for. They just are as they are, showing what they show and bringing into view more than words alone might do.

Using Ceremony

Most cultures honor significant emotions, such as grief following the death of a loved one or gratitude at times of thanksgiving, by marking them with ceremonies or forms of symbolic representation. We can honor our feelings of pain for the world by developing our own ceremonies to mark them.

TRY THIS: A PERSONAL CAIRN OF MOURNING

What is being lost in our world that you mourn for? While holding this question in your mind and heart, go for a walk and look for an object that represents something precious our world is losing. Find a special place to lay this object to rest as a way of acknowledging the loss and the grief you feel. You can return to this special place to mark other losses, over time building a small pile, or cairn, of objects symbolizing what is being lost.

This cairn of mourning is a marker affirming that you've noticed and that you care.

If we felt the pain of loss each time an ecosystem was destroyed, a species wiped out, or a child killed by war or starvation, we wouldn't be able to continue living the way we do. It would tear us apart inside. The losses continue because they aren't registered, they aren't marked, they aren't seen as important. By choosing to honor the pain of loss rather than discounting it, we break the spell that numbs us to the dismantling of our world.

The Power of Conversations

An expression we've often heard is "I didn't realize I felt so strongly until I heard myself say that." By speaking our concerns and giving voice to our feelings, we make them more visible not just to others but also to ourselves. The more we draw issues into the open, the more inclined we become to tackle them. As bestselling author Margaret Wheatley observes: "Many large-scale change efforts — some of which won the Nobel Peace Prize — began with the simple but courageous act of friends talking to one another about their fears and dreams. In reviewing a number of these efforts, I always found a phrase: 'Some friends and I started talking...'"[14]

When people reveal to us their anguish about world conditions, our response powerfully influences the way the conversation develops. Recognizing the sharing of pain for the world as a crucial communication, we can honor its expression by giving our full attention. Rather than attempting to fix feelings of distress, we accept their validity and significance. Doing this is in itself an act for the Great Turning.

Workshops and Practice Groups

A particular kind of magic happens when the Work That Reconnects is applied in a group setting. It is tremendously reassuring to find we are not alone in feeling pain for our world. To name it as normal is itself a healing act. See the resources section in the back of the book to find out about workshops near you.

Another way to tread the journey of the spiral with others is to form a support, practice, or study group. It can be based on reading this book together or drawing on the other resources listed on pages 253–55.

PAIN FOR THE WORLD AS A CALL TO ADVENTURE

There comes a point in many adventure stories when the main characters have seen the true nature of the crisis they face and it is far worse that they'd previously imagined. It is not just their own lives that are at stake but also a community they belong to, a value they cherish, a cause larger than themselves. With the limited tools and resources available to them, it might seem that they don't stand much chance. To have any hope at all, they need to step out onto new ground in a quest to find the allies, understandings, and tools that will help.

We will have moments when the penny really drops that our world is in grave danger. When facing a challenge far beyond what we might normally think ourselves capable of dealing with, we need to move past the familiar and learn the art of seeing with new eyes.

We use the phrase "seeing with new eyes" — the third stage in the spiral of the Work That Reconnects — in two different ways. The first refers more generally to the transformative shift that can happen when we find a different way of looking at something. Have you ever had the experience of feeling defeated or blocked in an area, but then discovering that things were not as they seemed as you opened to a new understanding? It is through "aha moments" like these that breakthroughs can happen. With this next part of the spiral, we invite in an openness to unfamiliar perspectives and a heightened willingness to step onto new ground. This may involve noticing what comes through sensory channels outside our everyday experience, including our imagination and our body. We might also be more willing to listen to people with views quite different from ours, as well as to the ancestors and future beings and the more-than-human world.

The second way we view this part of the spiral relates to a shift in our understanding of reality, a change in worldview from the hyperindividualism dominant in Western industrialized societies. It involves turning toward a holistic, ecological view that draws from both contemporary science and indigenous Earth wisdom traditions in many parts of our world. The next section of our journey explores this shift by focusing on four particular aspects. We like to think of these as the four discoveries: a wider sense of self, a different kind of power, a richer experience of community, and a larger view of time.

PART TWO

Seeing with New Eyes

CHAPTER FIVE

A Wider Sense of Self

In a fictional future on a faraway world, the indigenous Na'vi have developed such a richly satisfying life of connectedness that they can't be bought off. Preserving the beauty and vitality of their world is more important to them than anything a materialistic society can offer. This story, familiar to anyone who has seen the film *Avatar*, is set on a fertile moon called Pandora. The events the film portrays are recognizable on our planet too. Beautiful forests are being torn down to make way for open-pit mines, while corporate-backed mercenaries crush opposition from the local populations.

At the start of an epic adventure, the central characters often appear poorly equipped for the challenges they face. Whether it is David up against Goliath or Frodo bearing the ring, the hero of the story seems hopelessly underpowered for the role he or she steps forward to play. We can feel like that too. "Who am I to take on the problems of the world?" we might ask. Yet our view of what we're capable of is linked to our sense of who and what we are. Discovering hidden depths to our identity opens up a whole new set of possibilities. In *Avatar*, Jake Sully makes this journey, starting out as a loyal servant of the corporation and gradually finding himself a defender of the world he has joined. This chapter explores how we can experience a comparable transformation, in which the emergence of a wider sense of self powerfully enhances our ability to contribute to our world.

TRANSFORMING SELFISHNESS

People often despair about the problem of human selfishness. If, as some believe, we are genetically programmed to put our comfort and convenience before the interests of others or our world, then the outlook is pretty dismal. But what if we could transform the expression of selfishness by widening and deepening the self for whom we act? Arne Næss issued this invitation when he wrote: "Unhappily, the extensive moralizing within the ecological movement has given the public the false impression that they are being asked to make a sacrifice — to show more responsibility, more concern and a nicer moral standard. But all of that would flow naturally and easily if the self were widened and deepened so that protection of nature was felt and perceived as protection of our very selves."[1]

Many of us may remember special moments when we have felt unusually connected to the world around us. Perhaps we were enthralled by a glimpse of great beauty, or we witnessed an event so remarkable it took our breath away. "I have been a recipient of an unmistakable grace," wrote philosopher and author Sam Keen, describing the time he spent out in the woods watching two bobcats sitting on a log singing in the night.[2] These cherished moments, which some would describe as spiritual experiences, nourish us. They pull us out of preoccupation with our personal details and into a larger, more mysterious, and magical experience of being alive.

By inviting in these experiences of interconnectedness, we can enhance our sense of belonging to our world. This mode of being widens and deepens our sense of who we are.

DIFFERENT VIEWS OF THE SELF

When author and filmmaker Helena Norberg-Hodge first visited the mountainous region of Ladakh in northern India, she was struck by how happy the people were.[3] Even though the physical conditions were challenging, with villages cut off by snow for many months each year, the villagers had a rich and satisfying way of life that met their basic needs well. Though the villages she visited had no electricity, the villagers' highly developed cooperative culture ensured that everyone had enough food, clothes, and shelter. There was hardly any crime, depression was rare, and suicides were virtually unheard of. That was in the mid-1970s.

Since then, the picture has changed. With huge billboard advertisements in the towns glamorizing the path to material prosperity, many of the younger generation have left in search of employment. While some now have motorcycles and digital devices, there have also been suicides, even among school-age children. Previously, success and failure had largely been shared experiences; villagers worked together to succeed in tasks like bringing in a harvest. The modern consumer culture, however, has brought in its wake the rat race that divides people into winners and losers, with constant pressure to come out on top and to get hold of the fancy new products they've learned how to want.

The extreme individualism of industrialized countries is so well established that some regard it as our natural state. Seeing it emerge as a fairly new phenomenon in Ladakh is a reminder that the desire to put ourselves first isn't a fixed feature of human nature. Rather, it is a product of a particular way of understanding and experiencing the self that arises from the hyperindividualism developed over the past five centuries in Western culture.

There are many ways of viewing the self. The concept of a separate self, a discrete entity detached from other people and the world, is just one approach. This separate individual self, or "ego," as it is sometimes called, has a clear outer boundary. This isn't all of us, though, as Carl Jung once wrote: "It is a remarkable fact that a life lived entirely from the ego is dull not only for the person himself but for all concerned."[4]

The following exercise offers an opportunity to explore who else we are besides just our ego self.

Try This: Tell Me, Who Are You?

This process can be done alone or with a partner.

By yourself: Imagine encountering a stranger who's keen to get to know you and who asks, "Tell me, who are you?" Write your reply in a notebook. When you finish a response, imagine being asked the same question again. And again. See if you can respond a different way each time, writing down whatever answer feels right, and then repeating the process. Aim for at least ten different responses. If you're feeling curious, see if you can fill a whole page.

With a partner: Take turns, with each of you taking two or three minutes in each role, one asking, "Tell me, who are you?" and the other replying. Repeat the process again and again, allowing whatever words come up. Rest assured that the answers will probably be different every time.

When doing this exercise, people are often surprised by how many different ways they see themselves. Even when we describe ourselves as distinct individuals, our sense of identity also involves a connected self that emerges from our relationships, contexts, and communities. As different aspects of who we are grow more or less pronounced, our sense of self changes over time.

One workshop participant, Tom, experienced the process this way: Throughout his twenties, Tom had seen his main purpose as enjoying life. He wasn't much interested in world issues. He had enough going on in his own life. In his early thirties he got married, and a short while later, his wife became pregnant. The day he saw the ultrasound scan, his whole view of who and what he was changed. There in front of him was proof that he was about to become a father.

For the first time in his life, what was going to happen after he died started to become important to him. He began to think in a time frame that extended beyond his own life. Tom no longer thought of himself as only a separate individual. He experienced himself to be part of something larger — a family.

Tom discovered a wider and deeper sense of self. He didn't stop being an individual; rather, his personal identity became rooted in something larger — in this case, the shared identity of his family.

Widening Circles of Self

When Tom became a parent, he grew closer to other members of his and his wife's family, particularly those who helped in the care of their child. In his twenties, he had lost the sense of being part of an extended family. While he had felt independent and free to follow his own desires, at times he had felt very alone and lacking in a meaningful sense of purpose. Reconnecting with his family restored a sense of belonging. Feeling grateful for the help he was getting, he experienced a sense of kinship that left him wanting to help other family members too.

We don't think of it as altruism when family members help one another or when parents make sacrifices for the benefit of their children. It is viewed as entirely normal, even taken for granted. In a similar way, when we feel part of a team or a circle of friends, it is natural to give our support without charging by the hour or asking what's in it for us. Arne Næss comments on this cooperative tendency: "Early in life, the social self is sufficiently developed so that we do not prefer to eat a big cake alone. We share the cake with our family and friends. We identify with these people sufficiently to see our joy in their joy, and to see our disappointment in theirs."[5]

When we identify with something larger than ourselves, whether that be our family, a circle of friends, a team, or a community, that becomes part of who we are. There is so much more to us than just a separate self; our connected self is based on recognizing that we are part of many larger circles.

In the course of a day, we move among these different expressions of identity. We might inhabit our family self when attending to relatives, our team self when playing a role at work, our social self when enjoying a walk home with a friend. If we have a patriotic moment, that's an expression of our national identity. If we feel a sense of belonging when viewing a photo of Earth from space, that reflects the planetary dimension of our connected self. Our sense of rootedness comes from experiencing these more encompassing circles of our identity (see figure 10).

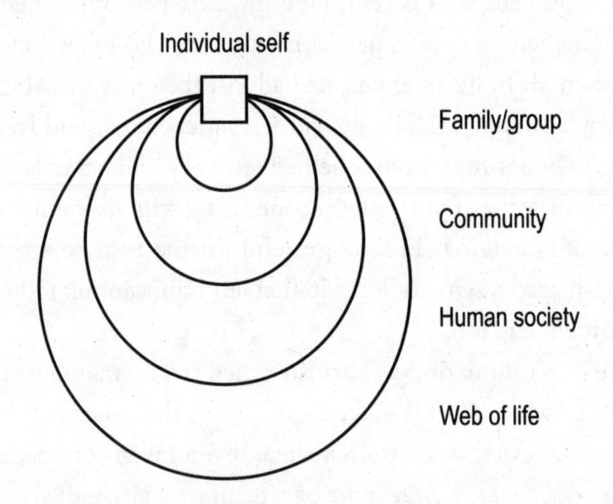

Figure 10. We are part of widening circles.

How we define our self-interest depends on which dimensions of self we are identifying with most strongly in that moment. Psychologist Marilynn Brewer writes, "When the definition of self changes, the meaning of self-interest and self-serving motivations changes accordingly."[6]

The distinction often made between selfishness and altruism is therefore misleading. It is based on a split between self and other, presenting us with a choice between helping ourselves (selfishness) and helping others (altruism). When we consider the connected self, we recognize this choice as nonsense. It is from our connected selves that much of what people most value in life emerges, including love, friendship, loyalty, trust, relationship, belonging, purpose, gratitude, spirituality, mutual aid, and meaning.

The philosopher Immanuel Kant made a distinction between "moral acts" and "beautiful acts."[7] We tend to perform moral acts out of a sense of duty or obligation. In contrast, we perform beautiful acts when we do what is morally right because it is attractive to us, the action motivated more by desire than duty. When our connected sense of self is well developed, we are more often drawn to beautiful acts. When we lose our sense of felt connectedness, we miss out on this sort of beauty, with tragic consequences.

When people lose their sense of belonging to larger circles, they lose not only the motivation to act for their communities and environment but also valuable sources of support and resilience. Alongside the erosion of extended family and community networks, the rate of depression in industrialized countries has been steadily rising for more than fifty years. It has now reached such epidemic proportions that one in two of us is likely to suffer a significant depressive episode at some point in our lives.[8]

What we see here is how personal well-being, community well-being, and planetary well-being are linked to the way we view our self. The extreme individualism of our culture is harmful at all three levels. To promote the recovery of our world and the healing of our communities while also leading lives that are rich and satisfying, we need to embody a larger story of who and what we are.

Interconnectedness Is Not about Merging

Experiencing the wider identity of our connected self does not mean losing our individuality. Quite the opposite: it is through finding and playing our unique role within a community that we feel more strongly part of it. This recognition is an important shift away from the idea that cohesion and order depend on everyone thinking and acting the same way. That would simply lead to a mob mentality, in which our uniqueness is submerged and our autonomy abandoned.

For a complex system to self-organize and function well, it requires both the integration and the differentiation of its parts. Complex organisms, such as our bodies, require many different types of cells, and a resilient ecosystem requires high levels of biodiversity. Compare that with monocrop agriculture, in which rows and rows of the same plant growing the same way might look ordered but are dependent on chemical inputs and are vulnerable to changing conditions. It is similar with human beings. We don't need to be like the Borg in *Star Trek*, robbed of free will and individuality, to experience ourselves as connected parts of a larger whole. The courage to listen to our conscience and live our own truth is integral to joining, rather than merging with, the larger circles of life we belong to.

When you fall in love, you feel incredibly bonded with your loved one and at the same time more uniquely yourself, different from anyone else in the world. The experience of connectedness with a healthy community has similar qualities: it brings out our latent distinctive gifts.

Restoring and "Restorying" Connectedness

Rather than viewing our self as a fixed thing with characteristics that can't be changed, we can think of our self as a flow of becoming. The static view of self is similar to a picture hanging on a wall, something that is set in a particular way and that resists transformation. Whenever we have thoughts like "I'm not the sort of person who . . . ," we're painting a similarly static picture of ourselves. An alternative view is to think of each moment as similar to a frame in a movie. If something isn't in the frame right now, that doesn't mean it won't be later. This perspective invites a greater sense of possibility.

We participate in many flows of becoming, from our own life to the lives of our family, community, and world. Each flow can be thought of as a story that moves through the players in it. With our individual self, the plot revolves around our personal adventures, gains, and losses. With our family self, the narrative can be traced back through our ancestors and extends into future generations. If we identify with a particular cultural or religious community, we are part of its story as well.

Arne Næss introduced the term *ecological self* to describe the wider sense of identity that arises when our self-interest includes the natural world.[9] When that happens, we move into a much larger story of who and what we are. Recognizing ourselves as part of the living body of Earth opens us to a great source of strength. The expression "Act your age" takes on a different meaning when we see ourselves as part of an amazing flow of life that started on this planet more than three and a half billion years ago. We come from an unbroken lineage that has survived five mass extinctions. Life has a potent creative energy and manifests a powerful desire to continue. When we align ourselves with the well-being of our world, we allow that desire and creative energy to act through us. When asked how he handles despair, rain forest activist John Seed replied: "I try to remember that it's not me, John Seed, trying to protect the rainforest. Rather, I am part of the rainforest protecting itself. I am that part of the rainforest recently emerged into human thinking."[10]

The understanding that we are an intrinsic part of the living Earth lies at the heart of indigenous belief systems around the world. Describing the traditional view of the world he grew up with in Malawi, theologian Harvey Sindima writes, "We live in the web of life in reciprocity with people, other creatures, and the earth, recognizing that they are part of us and we are part of them."[11]

Our Larger Selves Feel through Us

When we see ourselves solely as separate individuals, we think of our feelings as arising within us and as understandable purely in terms of our own story. The approach of family therapy, based on understanding the family as a living system, takes a different view. A system is a whole that is more than the sum of its parts; what makes a family more than just a collection of individuals is the way it functions as a whole. A family does this by feeling and acting through its members.

When a family is functioning well, it rallies around and supports individual members through times of hardship. If there is a crisis, such as a family member becoming seriously ill, the emotions of alarm and concern will pass through the family, often transmitted from one person to another, and activate a collective response. This same process also happens with other human systems, such as cohesive teams or communities.

We can look at our pain for the world in a similar way: it is the world system, or Gaia, feeling through us. The idea of being one with all life might seem abstract or like an advanced spiritual state attainable only after many years of meditation. Yet when we're moved to tears by whichever wounds of our world hurt us most, we experience, directly, our interconnectedness with the living Earth.

Different Stories of Evolution

According to neo-Darwinists, evolution is a product of fierce competition, with each species pitted against the others in a vicious battle to survive. A quite different perspective now accepted by mainstream science is called *endosymbiotic theory*. This theory proposes that important steps in our evolution have occurred through cooperation between species, even to the point of separate organisms joining together to create entirely new forms of life. As its leading proponents, Lynn Margulis and Dorion Sagan, write, "Life did not take over the globe by combat, but by networking."[12]

If we trace the development of life from its earliest beginnings, we see a recurring pattern of smaller parts coming together to form larger integrated wholes. Initially there were only single-celled organisms. At some point in our evolutionary history came the remarkable transition from separate cells doing their own thing to organized groups of cells acting as single organisms. A further evolutionary jump came with the progression from simple multicelled organisms to complex forms of life containing different organs. Another major step came with the first social organisms, as insects like ants and bees formed complex communities that functioned as integrated systems. Human beings express this social tendency too, though at a higher level of complexity, which allows individuals to make their own choices while at the same time being part of a larger community.

Throughout human history we have followed this same evolutionary pattern of smaller parts coming together to form larger complex wholes. For hundreds of thousands of years, we existed as hunter-gatherers in small tribal groups highly responsive to changing conditions. With the development of agriculture, groups settled into larger village communities. Over time, villages joined together within clans or tribes, and these larger regional groupings formed alliances that became nations. Again and again peoples previously divided have made common cause and found a larger shared identity.

Crisis is one of the factors that can motivate people to pull together. It can also have the opposite effect, triggering the collapse of communities and the unraveling of shared identities. With the planetary emergency we face, we are in danger of falling apart into factions fighting for what is left in a world we have depleted. But another possibility is that this crisis can become a turning point, the very danger moving us toward the next evolutionary leap.

THE EMERGENCE OF CONNECTED CONSCIOUSNESS

Something very interesting occurs when a group of jazz musicians improvise together. A number of separate individuals, all making their own decisions, act together as a whole. As the music flows, any of the musicians can take the solo spot, that leading role gliding seamlessly between the players. Who decides when the piano or trumpet player should come forward? It isn't just the person playing that instrument, for the others have already stepped back just a little to create an opening. Two levels of thinking are happening at the same time here: choices are made from moment to moment both by the group as a whole and by the individuals within it.

When people coordinate their actions through a collective thinking process, we can think of this as *distributed intelligence*. No one person is in charge; the players act freely while being guided by their intention to serve the purpose of the group. For musicians to improvise together, they need to listen very attentively, expressing their individuality in a way that contributes to the overall sound. When they tune in to the group and become connected with it, it is as though the music plays itself through them.

A key feature of distributed intelligence is that no one part has to have the whole answer. Rather, the intelligence of the whole emerges through the actions and interactions of its parts. In a creative team, an idea may arise in conversation, then be added to and refined by other team members, its development shaped by everyone present. What allows a team to gel is a shift in identification, so that people identify with and act for the team rather than just themselves.

Could the next leap in evolution arise out of a shift in identification, in which we shed the story of battling for supremacy and move instead to playing our role as part of the larger team of life on Earth? Could the creativity and survival instinct of humanity as a whole, or even of life as a whole, act through us? Here connected consciousness stems from a widening of our self-interest, where we are guided by the intention to act for the well-being of all life. Within Buddhism, that intention is known as *bodhichitta*. Bodhichitta moves our focus from personal well-being to collective well-being.

We stand at an evolutionary crossroads, and collectively we could turn either way. Our own choices are part of that turning. We can choose, to borrow a phrase from *Star Trek*, the "prime directive" of our lives. When our central organizing priority becomes the well-being of all life, then what happens through us is the recovery of our world. While we don't know how much this recovery process will be able to achieve, we can know that we give it our support.

The Shambhala Warrior Prophecy

Our decisions steer the flow not only of our lives but also of the unfolding stories we participate in. An epic story that has inspired both of us comes from the Tibetan Buddhist tradition: a twelve-centuries-old prophecy known as the Coming of the Kingdom of Shambhala. The heroes of this story, the Shambhala warriors, seek to save the world from destruction through the kind of power they know best. Inspired by the teachings of the Buddha, it is at root a story about the spiritual power of compassion, on the one hand, and on the other, of insight into the interconnectedness of all things.

Here is a version of the prophecy as it was given to Joanna

in the first days of 1980 by her dear friend and teacher Drugu Choegyal Rinpoche of Tashi Jong, a Tibetan community in exile in Northwest India. Though called a prophecy, the telling does not make clear whether the Shambhala warriors ultimately win the day over the barbarian forces.

> There comes a time when all life on Earth is in danger. At that time great powers have arisen, barbarian powers. Although they waste their wealth in preparations to annihilate each other, these forces have much in common. Among the things they have in common are weapons of unfathomable destructive power and technologies that lay waste to the world. It is just at this point, when the future of all beings hangs by the frailest of threads, that the kingdom of Shambhala emerges.
>
> Now, you cannot go there because it is not a place. It exists in the hearts and minds of the Shambhala warriors. And you can't tell whether someone is a Shambhala warrior just by looking at them, because they wear no uniforms or insignia. They have no banners to show which side they are on, no barricades on which to climb to threaten the enemy or behind which to rest or regroup. The Shambhala warriors don't have any home turf; there is only the terrain of the barbarian powers for them to move on and use.
>
> Now, the time comes when great courage is required of the Shambhala warriors — moral courage and physical courage. That is because they are going right into the heart of the barbarians' power to dismantle their weapons — weapons in every sense of the word. As they make their way into the pits and citadels where the armaments are made, so are they also entering the corridors of power where the decisions are made.
>
> Now heed this: the Shambhala warriors know these weapons *can* be dismantled. That is because they are *manomaya*, meaning "mind made." Made by the human mind, they can

be unmade by the human mind. The dangers facing us are not fashioned by some satanic deity, or by an evil extraterrestrial force, or by some immutable, preordained fate. These dangers are created by our relationships, our habits, our choices.

"So now is the time," said Choegyal Rinpoche, "for the Shambhala warriors to go into training."

"How do they train?" Joanna asked.

"They train in the use of two implements or tools," he said. And when she asked, "What are they?," he held up his hands the way the lamas hold up the ritual objects — the dorje and bell — in the great monastic dances of his people. "One," he said, "is compassion. The other is insight into the radical interdependence of all things.

"You need both," he said. "You need compassion because it provides the fuel to move you out to where you need to be in order to do what you need to do. It means not being afraid of the suffering of your world. That tool is very hot. By itself, it is so hot, it can burn you out.

"So you need the other tool too, the insight that reveals the interbeing of all that is. When that dawns in you, then you know that this is not a battle between good people and bad people, but that the line between good and evil runs through the landscape of every human heart. And you know that we are so interwoven in the web of life that even our smallest acts have repercussions that ripple through the web of life beyond our capacity to discern.

"But that knowing is kind of cool; it can seem at times a bit abstract. That's why it needs the heat of compassion."

That is the essence of the prophecy. If you have watched Tibetan monks or nuns chanting, you've probably seen their hands making gestures, or mudras, at the same time. Their hands may well be dancing the interplay between compassion and insight, which is there for each of us to embody in our own way.

Applying the Prophecy in Our Lives

When sharing this prophecy in workshops and courses, we've been struck by how many people experience it as a powerful calling to step into a role much needed in our time. Ociane, moved to tears after watching a video of Joanna telling the tale, wrote: "The story was like a confirmation of what I've been seeing with my eyes and feeling in my bones. It was like realizing I've been knowing deep inside where I belong and what I have to do."

If we're inspired by a story and want to strengthen its impact in our lives, a practice we can use is to experiment with the sentence starter "What if..." For example, while recognizing its roots in a culture and time period different from our own, what if this was also a story about us, here, now? What if it were a story about you?

When we take this approach, the prophecy prompts us to recast our narrative about who we are and what role we are called to play in this time. Each "What if..." invites us to try out a different way of considering the story. We invite you to experiment with this practice and see what rings true for you. Here are some ways we find helpful to experience and receive the prophecy.

At a time of great danger, the kingdom of Shambhala emerges in the hearts and minds of the Shambhala warriors. What if we were to think of the kingdom of Shambhala as a deep knowing in our hearts and minds that we belong to this world, that it is the larger country we are from? If we see ourselves as part of the nation of life on Earth, then, rather than just belonging to this country or that, we recognize that we are a planet people who share the same home. We might look at an image of our Earth taken from space or at the land around us wherever we are and know in our bones that this is where we belong, this is where we're from.

The word *warrior* doesn't refer only to people engaged in warfare or those who are good at fighting. It also means someone who stands up for a cause they have their heart in. We can view Shambhala warriors as people fiercely loyal to life on Earth who are determined to play their part in its defense.

The mission of these warriors is to dismantle the weapons that so gravely threaten our world. What if we were to see weapons not just in terms of armaments but also, more broadly, as anything that has harmful impact? For example, a gun is an intentional weapon — it is designed to cause harm. Yet, surprisingly, more people around the world are killed each year by cars than by guns.[13] That's even before we start counting the impact of air pollution, carbon emissions, and the ecological costs of road building. Perhaps the role of Shambhala warriors also includes dismantling our reliance on cars. We might redesign the way we live to reduce our need for travel and, for the traveling we do, turn toward less harmful methods.

In 2009, the Black Saturday wildfires in Australia killed more than 170 people. They also released energy equivalent to fifteen hundred atomic bombs the size of the one dropped on Hiroshima.[14] Australia's bushfire season in 2019–20 burned more than forty times as much ground as the Black Saturday fires and killed more than a billion wild animals.[15] The vast quantities of greenhouse gases released each year set us on a path that makes wildfires like these more likely. Our highly carbonized lifestyles and society act as weapons as much as our missiles and tanks. What if we were to see our mission as one of dismantling our dependence on fossil fuels?

If a weapon is something that causes harm, then that includes not just armaments but also the thinking that leads us into battle as a way of responding to conflict. What if dismantling weapons also included addressing the style of thinking that separates people into "us" and "them," that creates enemies and then stokes the enmities that turn us against one another?

Continuing the theme of exploring the impact of our view of reality, which is the more deadly weapon in the image (see figure 11) — the pickax or the thinking that leads someone to use it in a harmful way?

Figure 11. Thoughts with harmful impact act as weapons.

Perhaps the most harmful beliefs are those that dismiss the power and significance of our behaviors. They make it easy to discount both the harmful and the beneficial consequences of our actions, blocking the motivation to change. We need to dismantle the thinking styles that make it difficult to recognize our influence. Turning toward a more empowering view is therefore the focus of our next chapter.

CHAPTER SIX

A Different Kind of Power

From our wider identity as part of the living Earth comes a strong urge to act. Yet while pain for the world alerts us to the urgency of our situation, we may still view this planetary crisis as beyond our power to affect. In a survey of two thousand adults, Britain's Mental Health Foundation found that a feeling of powerlessness was by far the most common response to global issues.[1] Joe, a student blogging on the web, expressed his experience of helplessness:

> It seems to me that climate change is something only industrial and political leaders have any real power to solve. If we're honest, the idea that individuals like me can really contribute to a solution seems laughable. Am I wrong to think this way? Am I giving up?[2]

In seeing power as held only by those at the top of a hierarchy, Joe is expressing a view that is widespread and that undermines our capacity to act. When we see with new eyes, we discover a different way of perceiving and experiencing power. Before we move on to exploring this discovery, we need first to describe the old view of power and the problems it brings.

The Old View of Power

In the old story, power is based on a position of dominance or advantage over others, a position that secures privileged access to resources and influence. This type of power is about getting your way and having others do what you want. The bottom line here is being able to win in a conflict; the more you can beat others, the higher you rise. We refer to this type of power as *power over*. Let's see where it leads.

Feelings of Powerlessness Are Widespread

Power over is based on a win-lose model; in the contest to come out on top, most of us end up losing. The position of being "in power" is such a small space that only a few people can fit there, leaving many feeling, as Joe the blogger expressed, that "the idea that individuals like me can really contribute to a solution seems laughable."

While we may be able to get our own way in some areas of our personal life, there is a limit to what we see as within our power. Because global issues are usually considered far beyond this limit, we may view thinking about them as a waste of time. For example, an expression we've often heard is "There's no point worrying about things you can't change, and you can't change the world."

Power Is Viewed as a Commodity

What gives a person a position of advantage over others? It is having something that others do not — whether it be money, weapons, material resources, contacts, or information. When information gives advantage, it becomes a tradable asset; as a result, valuable knowledge gets hidden away, and public awareness is held back by secrecy.

Even votes in elections are thought of as buyable, with campaign funds dependent on huge donations from vested interest groups who expect favors in return. The power to shape the direction of our society has become a commodity to be bought and sold. When power is a possession to be held, defended, and accumulated, it becomes increasingly removed from the hands of ordinary people.

Power Generates Conflict

Power over is essentially oppositional, because gaining it involves taking it away from others. To rise over others, either as an individual or as a group, you need to push others down; to get in power, you need to push others out. As those pushed down and out are left with resentment, those in power then need to keep tabs on the opposition and stop them from becoming powerful enough to present a threat.

Fear is intrinsic to this model of power. Even if and when you are on top, you have to be vigilant, lest you lose the upper hand. In the struggle to stay on top, ruthlessness and dishonesty have become so common that the link between power and corruption is often seen as inevitable.

Dominance gives privileged access to resources, and to maintain dominance, huge amounts are spent on being "strong," that is, able to win in a fight. In 2020 the global arms expenditure was nearly $2 trillion.[3] For perspective, spending just 1 percent of this amount each year could be enough to address world hunger.[4]

Power Fosters Mental Rigidity

When displays of strength are seen as important, changing one's mind is viewed as "giving in," as a sign of weakness. In political discussions, winning is valued more highly than deepening understanding. This standpoint blocks openness to new information and stifles the flexibility needed to deal with changing circumstances.

Power Becomes Suspect

When completing the sentence "Powerful people tend to be...," our workshop participants often reveal mixed feelings about becoming powerful, identifying both appealing and unappealing qualities. While they view powerful people as passionate, clear, determined, and brave, they also see them as more likely to be lonely, stabbed in the back, dishonest, and disliked.

This mixed picture presents a dilemma for those who want to find the power to make a difference in the world but not enter a battleground where they are likely to become distrusted, lonely, or corrupted. Suspicion of power leads people to be reluctant to act authoritatively. Many have become disillusioned with mainstream politics: witness the low turnout at elections in many democracies. Fortunately, the power-over model isn't the only way to understand power. When we see with new eyes, a more attractive and capacity-building alternative comes into view.

A New Story of Power

The word *power* comes from the Latin *potēre*, meaning "to be able." The kind of power we will now focus on is not about dominating others but about being able to address the mess we're in. Rather than being based on how much stuff or status we have, power in this view is rooted in insights and practices, in strengths and relationships, in compassion and connection with the web of life.

One person who embraced the collaborative model of power was Nelson Mandela. In the early 1980s, the apartheid government of South Africa had a highly trained army with advanced weaponry and nuclear missiles. Mandela, representing the African National Congress (ANC), had been in prison for more than twenty years. While many feared it might take a civil war to end it, apartheid didn't end through victory in battle. Rather, the transformation came about through discussion and agreement. For that process to start, it needed, as Mandela put it, "Jaw, jaw, jaw, not war, war, war."

In his autobiography, *Long Walk to Freedom*, he describes coming to a decision to move this process forward while in solitary confinement:

> My solitude gave me a certain liberty, and I resolved to use it to do something I had been pondering for a long time: begin

> discussions with the government....This would be extremely sensitive. Both sides regarded discussions as a sign of weakness and betrayal. Neither would come to the table until the other made significant concessions....Someone from our side needed to take the first step.[5]

When we respond to a situation in a way that promotes healing and transformation, we are expressing power. Mandela's contributions to establishing a multiracial democracy in South Africa offer a wonderfully inspiring example.

Because Mandela didn't have the authorization of the ANC's organizing committee, beginning talks with the enemy could have been seen as a betrayal or selling out. Taking this first step for peace required courage, determination, and foresight. Inner strengths like these are often thought of as things some people just have and others don't. These qualities, however, are linked to skills we can develop and practices we can learn. Thinking of courage and determination as things we *do* rather than things we *have* helps us to develop these qualities. They emerge out of our engagement with actual situations and the dynamics that arise from our interactions. This approach is relational, and we call it *power with*.

1 + 1 = 2 and a Bit

The discussions Mandela undertook were effective because both sides recognized that they stood to lose by going to war and they would gain by finding a way to peace. They moved from a win-lose model of conflict to one aiming for a win-win outcome. The alternative to negotiations was likely to be a war in which both sides had much to lose.

Power with is based on *synergy*, where two or more parties working together bring results that would not have occurred if they had worked alone or in competition. Because something new and different emerges out of the interaction, we can think of it as "1 + 1 = 2 and a bit." This is another way of saying, "The whole is more than the sum of the parts."

Emergence and synergy lie right at the heart of power with. They generate new possibilities and capacities, adding a mystery element because we can never be certain how a situation will go just from looking at what we start out with. We can know the strength of copper and of tin yet still be surprised by how much stronger bronze is, which combines the two. The same thing can happen when we interact with others for a shared purpose. D. H. Lawrence wrote:

> Water is H_2O,
> Hydrogen two parts
> Oxygen one
> But there is also a third thing that makes it water
> And nobody knows what that is.[6]

One place we can experience synergy is in conversation. When both sides have the courage and willingness to explore new ground, talking and listening to each other can open a creative space from which new possibilities emerge. That's what happened in the negotiations between Mandela and F. W. de Klerk, the South African president at the time. This unlikely duo jointly won the Nobel Peace Prize in 1993 for their extraordinary feat of navigating toward a peaceful settlement.

EMERGENCE

While the conversations between Mandela and de Klerk played a pivotal role in bringing apartheid to an end, this historic change would not have happened without a much larger context of support. Within South Africa, people risked their lives daily to engage in the struggle for change. Around the world, millions of people played supporting roles by joining boycotts and campaigns. If we focus only on individual, separate activities, it is easy to dismiss each one by thinking, "That won't do much." To see the power of a step, we need to ask, "What is it part of?" An action that might seem inconsequential by itself adds to and interacts with other actions in ways that contribute to a much bigger picture of change.

Remember our examples of digital images and newspaper photos? When highly magnified, they appear as just a collection of tiny pixels or dots, but when, from a little distance, we see the image as a whole, the larger pattern comes into view. In a similar fashion, a bigger picture of change emerges out of many tiny separate actions and choices. This link between small steps and big changes opens up our power in an entirely new way. Each individual step doesn't have to make a big impact on its own, because we can understand that the benefit of an action may not be visible at the level at which that action is taken.

Shared visions, values, and purposes flow through and between people. Nelson Mandela was deeply committed to a vision for his country that many were holding; the power of that vision moved through him and was transmitted to others. This type of power can't be hoarded or held back by prison walls; it is like a kind of electricity that lights us up inside and inspires those around us. When a vision moves through us, it becomes expressed in what we do, how we are, and what we say. The alignment of these three creates a whole that is more than the sum of its parts. The words below, from Mandela's defense at his trial in the 1960s, mean so much more because of the actions that followed them:

> During my lifetime I have dedicated myself to this struggle of the African people. I have fought against white domination and I have fought against black domination. I have cherished the ideal of a democratic and free society in which all persons

> live together in harmony and with equal opportunities. It is an ideal which I hope to live for and to achieve. But if needs be, it is an ideal for which I am prepared to die.[7]

THE POWER OF EMERGENCE

The concept of power with contains hidden depths; so far we've described four aspects. First, there is the power of inner strengths we draw from when we engage with challenges and rise to the occasion. Second, there is the power arising out of cooperation with others. Third, there is the subtle power of small steps whose impact becomes evident only when we step back and see the larger picture they contribute to. And last, there is the energizing power of an inspiring vision that moves through us and strengthens us when we act for a purpose bigger than ourselves. All these are products of synergy and emergence; they come about when different elements interact to become a whole that is more than the sum of its parts.

At every level, from atoms and molecules to cells, organs, and organisms, complex wholes arise, bringing new capacities into existence. At each level, the whole acts through its parts to achieve more than we could ever imagine from examining the parts alone. So what new capacities emerge when groups of people act together to form larger complex social systems?

Our technologically advanced society has achieved wonders our ancestors could never have envisioned. We've put people on the moon, decoded DNA, and cured diseases. The problem is this collective level of power is also destroying our world. Countless seemingly innocent activities and choices are acting together to bring about the sixth mass extinction in our planet's history.

Seeing with new eyes, we recognize that we're not separate individuals in our own little bubbles but interconnected parts in a much larger story. A question that helps us develop this wider view is "What is happening through me?" Is the sixth mass extinction happening through us as a result of our habits, choices, and actions? By recognizing the ways we contribute to the unraveling of our world, we identify choice points at which we can turn toward its healing. The question "How can the Great Turning happen through me?" invites a different story to flow through us. This type of power happens through our choices, through what we say and do and are.

Not Needing to Know the Outcome

The concept of emergence is liberating because it frees us from the need to see the results of our actions. Many of our planet's problems, such as climate change, mass starvation, and habitat loss, are so much bigger than we are that it is easy to believe we are wasting our time trying to solve them. If we depend on seeing the positive results of our individual steps, we'll avoid challenges that seem beyond what we can visibly influence. Yet our actions take effect through such multiplicities of synergy that we can't trace their causal chain. Everything we do has ripples of influence extending far beyond what we can see.

When we face a problem, a single brain cell doesn't come up with a solution, though it can participate in one. The process of thinking happens at a level higher than just individual brain cells — it happens *through* them. Similarly, there's no way that we personally can fix the mess our world is in, but the process of healing and recovery at a planetary level can happen through us and through what we do. For this to occur, we need to play our part. That's where power with comes in.

The Helping Hand of Grace

All the individuals on a team may each be brilliant by themselves, but if they don't shift their story from personal success to team success, their net effectiveness will be greatly reduced. When people experience themselves as part of a group with a shared purpose, team spirit flows through them, and their central organizing principle changes. The guiding question shifts from "What can I gain?" to "What can I give?"

We can develop a similar team spirit with life. When we are guided by our willingness to find and play our part, we can feel as if we are acting not just alone but as part of a larger team of life that acts with us and through us. Since this team involves many other players, unsuspected allies can emerge at crucial moments; unseen helpers can remove obstacles we didn't even know were there. When we're guided by questions such as "What can I offer?" and "What can I give?," we might sometimes play the role of stepping out in front and at other times the role of ally giving support. Either way, we can think of the additional support we receive as a form of grace. Based on an interview with Joanna, this poem, edited into verse by Tom Atlee, founder of the Co-Intelligence Institute, expresses well the grace that comes from belonging to life:

When you act on behalf
of something greater than yourself,
you begin
to feel it acting through you
with a power that is greater than your own.

This is grace.

Today, as we take risks
for the sake of something greater
than our separate, individual lives,
we are feeling graced
by other beings and by Earth itself.

Those with whom and on whose behalf we act

give us strength
and eloquence
and staying power
we didn't know we had.

We just need to practice knowing that
and remembering that we are sustained
by each other
in the web of life.
Our true power comes as a gift, like grace,
because in truth it is sustained by others.

If we practice drawing on the wisdom
and beauty
and strengths
of our fellow humans
and our fellow species
we can go into any situation
and trust
that the courage and intelligence required
will be supplied.[8]

Power With in Action

Here are three ways we can open to the kind of power we've been describing. We can:

- hear our call to action and choose to answer it.
- understand *power* as a verb.
- draw on the strengths of others.

Hear Our Call to Action and Choose to Answer It

There will be times when we become alerted to an issue and experience an inner call to respond. Choosing to respond to that call empowers us. Once we take that first step, we start on a journey presenting us with situations that increase our capacity to respond. Strengths such as courage, determination, and creativity are drawn forth from us most when we rise to the challenges that evoke them. When we share our cause with others, allies appear; synergy occurs. For when we act for causes larger than ourselves, the larger community for whom we do this starts acting through us.

We can experience our call to action in many different ways. Sometimes the uncomfortable discrepancy of realizing that our behavior is out of step with our values motivates us. Our conscience calls, and when we step into integrity, more of who we are heads in the same direction. At other times our call is more of a powerful summoning. We just know, even if we're not sure how, that we need to be somewhere, do something, or contact a particular person.

If we think of ourselves only as separate individuals, then we understand these intuitive calls purely in personal terms. Recognizing ourselves as part of the larger web of life expands our view. Just as we can experience our pain for the world as the Earth crying within us, we can experience the Earth thinking within us as a guiding impulse pulling us in a particular direction. We can view this as *co-intelligence*, an ability to think and feel with our world.

Developing a sense of partnership with the Earth involves listening for guiding signals and taking them seriously when we hear them. Our friend John Seed, the Australian rain forest activist, teaches a simple way of listening for such inner guidance. He calls it "a letter from Gaia," and it can be a powerful way of hearing our call to action.

Try This: A Letter from Gaia

If the Earth could speak to us, what would it say? We can take a step toward finding out by imagining the Earth writing through us. On a blank sheet of paper, start a letter to yourself that begins:

Dear [your name]:
This is your mother Gaia writing...

Continue the letter with whatever words come naturally. Let your hand do the writing, don't think about it too much, and just allow the words to flow. Feel free to start this letter with different words if you prefer.

Understand *Power* as a Verb

The old view of power is based on having more of something than the competing party. Power with, however, is not a property or possession. It arises from what we *do* rather than what we *have*. The shift in perception from seeing power as a noun to seeing it as a verb has surprising potency. Here are two open sentences that invite an exploration of *power* as a verb.

Try This: Open Sentences That Empower

These open sentences can be used in self-reflection, for journaling, or in a partnered listening exercise.

- I empower myself by...
- What empowers me is...

When we have explored this exercise in workshops, people have described empowering themselves by remembering what's important, doing what really matters, experiencing emotions, exercising regularly, eating well, getting enough sleep, seeking out good company, meditating, paying attention to needs (their own and those of others), laughing, dancing, and singing. When looking at what empowers them, participants have often mentioned inspiring purposes, friends who encourage and support, and a sense of rootedness in life. *Power* as verb points us in a very different direction than the noun form does.

Draw on the Strengths of Others

One way we can empower ourselves is by drawing on the strengths of others. A wonderful example of this can be found in T.H. White's *The Sword in the Stone*, when he tells the story of King Arthur as a boy being tutored by Merlin.

> The wizard Merlin, as Arthur's tutor, schooled the boy in wisdom by turning him into various creatures and had him live, for brief periods of time, as a falcon, an ant, a wild goose, a badger, a carp in the castle moat....
>
> The time came when the new King of All England was to be chosen: it would be he who could draw the sword from the stone. All the famous knights, who came to the great tournament, went to the churchyard where the stone mysteriously stood, and tried mightily to yank out the sword that was embedded in it up to the hilt. Heaving and sweating, they competed to prove their superior strength. No deal; tug and curse as they might, the sword did not budge. When the disgruntled knights departed to return to their jousting, Arthur, who was just a teenager then, lingered behind, went up to the stone to try his own luck. Grasping the sword's handle he pulled with all his strength, until he was exhausted and drenched. The sword remained immobile. Glancing around, he saw in

> the shrubbery surrounding the churchyard the forms of those with whom he had lived and learned. There they were: badger, falcon, ant and the others. As he greeted them with his eyes, he opened again to the powers he had experienced in each of them — the industry, the cunning, the quick boldness, the perseverance. Knowing they were with him, he turned back to the stone and, breathing easy, drew forth from it the sword, as smooth as a knife from butter.[9]

When we draw on a sense of fellowship, belonging, and connection, it is as if we are remembering our root system. This is power with, which comes from the larger circle that we can draw on and which acts through us.

In his workshops, Chris sometimes asks people to remember a time when they did something that made a difference. It doesn't have to be anything grand, just something positive that might not have occurred otherwise. Then, in groups of three or four, he asks people to take turns telling their stories and also identifying what strengths helped them take these actions. After doing this, people often say, "Hearing you describe using this strength helps me recognize it in myself." When other people open to their strengths, it can help us open to ours as well. We can catch this type of power from one another.

Whenever you are struggling, remember the sword in the stone. Think of trying to pull it out. Then pause. Remember those who inspire you. Think of them around you and feel their strengths joining yours. Think of all who support and believe in you. Draw strength from them as well. Think of who and what you are acting for and feel their power acting through you too.

CHAPTER SEVEN

A Richer Experience of Community

There is an old folktale from Denmark about a meeting between two kings. "You see that tower?" said the first king to the second, pointing to a tall, highly fortified part of his castle. "In my kingdom, I can command any of my subjects to climb to the top and then jump to their deaths. Such is my power that all will obey." The second king, who was visiting, looked around him and then pointed to a small, humble dwelling nearby. "In my kingdom," he said, "I can knock on the door of a house like that, and in any town or village, I will be welcomed. Such is my power that I can stay overnight, sleeping well without any fear for my safety."

The first king had power over, and the second king had power with. When we follow the path of partnership, a different quality of relationship emerges and, with it, a richer experience of community. In this chapter we'll look at fellowship and community as forms of wealth that enrich our lives, strengthen our security, and give us a more stable foundation from which to act.

The Epidemic of Loneliness

The term *community* is often used to describe those living in a specific location or having something in common, such as a profession, activity, or ethnic background. But in modern urban environments, people can live in the same building yet still have no real connection to one another. Commenting on the widespread lack of social cohesion in Western society, psychiatrist M. Scott Peck describes his experience of growing up in a New York apartment block: "This building was the compact home for twenty-two families. I knew the last name of the family across the foyer. I never knew the first names of their children. I stepped foot in their apartment once in those seventeen years. I knew the last names of two other families in the building. I could not even address the remaining eighteen."[1]

The danger of being too comfortable, too self-sufficient, is that we lose any sense of needing one another. If each family has everything it needs, what reason do we have to knock on our neighbors' doors? Experiencing need prompts people to reach out and make contact. That is why self-help groups like Alcoholics Anonymous have become such fertile expressions of community and fellowship. Through painful experience, their members have learned the truth of the maxim "I can't, we can." Crisis becomes a turning point when it provokes us to reach out to others.

Where do we look to meet our needs for security and a satisfying life? In the story of Business as Usual, the central plot is about personal advancement through economic success; the assumption is that we meet our needs by getting more and better stuff. This story leads people to invest time, resources, and attention in their own little bubbles rather than in their relationships and communities. In the United States, for example, the proportion of people with no one to confide in has nearly tripled in recent decades.[2] A survey published just before the pandemic in 2020 found that more than 60 percent described themselves as lonely.[3] Behind the locked doors lies an epidemic of aloneness that extends throughout the industrialized world.

Networks of mutual support bring many benefits, including higher perceptions of well-being and self-rated health, less loneliness, lower levels of stress, less depression, improved child development, reduced crime rates, a reduced risk of heart attacks, and lower overall mortality.[4] Referred to as "social capital," the web of supportive relationships within a neighborhood is a form of wealth that improves the quality of our lives. Unfortunately, with the trend toward increased individualism and consumerism, this great treasure is in decline. It has taken a further severe knock with the impact of the pandemic. While there have been many inspiring examples of people stepping up to build networks of mutual support, the impact of Covid has also been to leave many people feeling more distant from one another.

A Different World Is Possible

Once in a while something occurs that sweeps away the isolation and mutual indifference so prevalent in modern society. Rebecca Solnit, in her book *A Paradise Built in Hell*, describes such a case:

> I landed in Halifax, Nova Scotia, shortly after a big hurricane tore up the city in October of 2003. The man in charge of taking me around told me about the hurricane — not about the winds that roared at more than a hundred miles an hour and tore up trees, roofs, and telephone poles, or about the seas that rose nearly ten feet, but about the neighbors. He spoke of the few days when everything was disrupted, and he lit up with happiness as he did so. In his neighborhood all the people had come out of their houses to speak with each other, aid each other, improvise a community kitchen, make sure the elders were okay, and spend time together, no longer strangers.[5]

Our human tendency to pull together in emergency situations, even to risk our lives helping others, is well documented. In her study of human responses to disasters, Solnit describes how such rising to the occasion is more common than many suspect and more satisfying than many might imagine. Referring to her own experience of an earthquake in the San Francisco Bay Area in 1989 and the gratifying engagement she noticed in herself and others afterward, she writes of feeling:

> that sense of immersion in the moment and solidarity with others caused by the rupture in everyday life, an emotion graver than happiness but deeply positive. We don't even have a language for this emotion, in which the wonderful comes wrapped in the terrible, joy in sorrow, courage in fear. We cannot welcome disaster, but we can value the responses, both practical and psychological.... Disasters provide an extraordinary window into social desire and possibility.[6]

When food appears reliably on our tables, we don't need to exercise our creativity or social intelligence to survive. It is different in a disaster. The closeness of danger activates our wits and our cooperative tendencies in ways that bring out new levels of aliveness and community. When the shops are flooded and the system is in disarray, the helping hand of a neighbor or an improvised soup kitchen offer more security than status or money. As we reach out to help one another, our lives become more meaningful and satisfying. We discover that we don't survive, or thrive, alone. That's what Helena Norberg-Hodge saw in the Ladakhi villagers of northern India (see chapter 5); knowing that their survival depended on the land and people around them, they experienced their interdependence as the very foundation of their reality.

What comes into view when we see with new eyes is this interdependence. There is no such thing as a "self-made person": while we play a role in making ourselves, we are also made by one another and our world. When hurricanes, floods, and earthquakes sweep away illusions of self-sufficiency, we're reminded how much we need one another, how much we depend not only on people but also on the larger web of life. We treat people with a different kind of respect when we consider that they might someday be pulling us out of the rubble. We treat the rest of life with a different kind of respect when we consider that without it, we wouldn't be here at all.

Familiar Bubbles Beginning to Burst

We don't have to wait for a natural disaster before opening to the rich experience of community described by Solnit and others. We see this opening in our workshops on a regular basis. As participants express their grief, dread, and outrage at the unraveling of our world, awareness grows that we are, all of us, living in a disaster zone. Hearing one another name the tragedy unfolding in our world reassures us that we're not alone in noticing. The expression of common concern builds community, as does our shared willingness to show up and respond.

In our workshops we often use an exercise called the Milling, in which participants move around the room and then stop to face each other in pairs. We invite them to consider the possibility that the person in front of them might become a victim of the unraveling we face. With the threat of environmentally linked cancers, nuclear warheads still poised for action, pandemic infections, and an increase in climate-related disasters, such a prospect is sadly plausible. We then ask participants to acknowledge that the person they face might make a crucial contribution to the healing of our world. This too is possible.

This process drops the veil of normality that screens out the very real perils we face. The bubble of Business as Usual dissolves for a moment. The exercise also recognizes that each of us can make a pivotal contribution to our world. We can never know whether our actions will have a decisive impact. What we can know is that by supporting one another, we make this possibility more likely.

At the moment, the impact of the unraveling is unevenly distributed. While climate-related disasters have wrecked the lives of many millions of people, millions more, from the comfort of their homes, still don't believe there is much of a problem.

As the unraveling of our world proceeds, it becomes more difficult to hide from. Unfortunately, shared awareness of a problem does not inevitably cause people to come together as a community of mutual support. When the danger is only vaguely sensed and not understood, this can lead to distrust, hostility, and scapegoating. In adversity, people can pull together or turn against one another, step outside their bubbles or retreat further into them.

With the Covid pandemic, we've seen both things happen. Against a background of fear and disruption, misinformation and conspiracy theories spread rapidly through social media, splitting families and friendships in an extreme polarization of views. At the same time, even though necessary distancing blocked so many aspects of community life, there have been inspiring examples of mutual aid, as people reached out to help neighbors they'd perhaps not even met before.

The way we understand power greatly influences which way we go. The more that people and nations apply a power-over model, the more they rely on force to maintain positions of advantage. This perspective fills the world with enemies against whom we must defend ourselves.

Four Levels of Community

The Shambhala warrior prophecy, recounted in chapter 5, describes a time of great danger when the hazards we face arise out of our relationships, habits, and priorities. The quest of the Shambhala warrior is to dismantle the mind-made weapons laying waste to our world.

We've already explored how the term *weapon* applies not only to armaments but also to destructive ways of thinking and acting. The pattern of thinking that divides humanity into "us" and "them" is something we can all play a role in dismantling. To do so, we use our two implements: compassion and insight into our interconnectivity with all life. These implements undo the thinking that creates enemies, leading us into a richer experience of community.

We can think of community as having different levels. Each progressively widens our sense of what we belong to, what we are nourished by, and what we act for. These levels are:

- groups we feel at home in.
- the wider community around us.
- the global community of humanity.
- the Earth community of life.

At each level, we can apply the implements of insight and compassion to dismantle the thinking that fragments our world and sets us against one another. The process of building community is self-reinforcing, since not only does it contribute to the healing of our world, but it also enhances the quality of our lives. Like the second king in the Danish folktale, when we feel welcomed rather than threatened by those around us, we sleep more soundly at night. Let us look at each of these community levels in turn.

Groups We Feel at Home In

A group we feel at home in is small enough that we know one another's names and share common interests, even a common purpose. Feeling at home in such a group doesn't always happen immediately; it can take time to build trust and a sense of ease. When we feel the bond of common cause and reciprocal support, we have a powerful setting for synergy.

We see this level of community at work in adventure stories in which a deeper purpose acts through a small group of central characters in a way that forges extraordinary loyalty among them. The bond among Harry, Hermione, and Ron in the Harry Potter stories grows out of their shared response to the grave danger they recognize. In *The Lord of the Rings*, Frodo acts with allies in the fellowship of the ring to accomplish their mission, and the lengths his friends go to for one another create a tough, enduring sense of community. The same can happen in our lives when we act with others toward shared goals.

A group we feel at home in can support us through remarkable personal transformations. Our interactions change when we feel safe enough to let go of defensiveness, allowing us to become more open to one another and to life. Ian, a colleague, described his experience in a group committed to supporting its members in offering their best contribution to the world: "I had found a place where I could make a contribution simply by showing up and showing care for those who also showed up. Slowly I found my voice in the group. I felt supported; it was like finding nourishing soil in which to grow."

As a lone voice, we can feel drowned out by the constant broadcast of commercial reality and swept along by the rush of Business as Usual. But a kind of magic can happen in groups like the one Ian described. The fellowship generated anchors and nourishes us. Circles of fellowship create space for a different story to be heard, spoken, and lived. By providing a protected space to share our concerns and sprout new responses, they serve as seedbeds for the Great Turning.

Seeing with new eyes involves recognizing a story much bigger than our personal dramas. Doing so fosters a different type of interpersonal economics, one with increased generosity and understanding. Issues such as who's getting the best deal or who has the most status fade in comparison with what we can achieve together. Acting as a group for the Great Turning elevates our friendships and graces them with new beauty. Such groups powerfully support our ability to bring healing and transformation to our world. As the quote widely attributed to anthropologist Margaret Mead affirms: "Never doubt that a small group of thoughtful, committed citizens can change the world. Indeed, it is the only thing that ever has."[7]

We need groups like this now and will need them all the more in the coming years. They give us a foundation of resilience that helps us adapt to changing circumstances, recover from setbacks, and find strength in times of adversity. When conditions are difficult, having a trusted gang around us both to draw from and to give to can make all the difference. Doris Haddock, the activist fondly known as Granny D, was ninety-eight years old when she gave a talk in Philadelphia describing how this sort of mutual support transformed her experience of the Great Depression: "Maybe we were hungry sometimes, but did we starve? No, because we had our friends and family and the earth to sustain us. . . . We were fountains of creativity. We were fountains of friendship to our neighbors. As a nation, we were a mighty river of mutual support."[8]

The immediate circle in which we feel most at home is just the first rung of community. It is easy to build community with those who are like us and share our point of view. We start with what is close and nourishing, but that is just the beginning. To dismantle the weapons laying waste to our world, we need to extend our community beyond this.

The Wider Community around Us

In 1958 Dr. A. T. Ariyaratne was a young science teacher at a high school in Sri Lanka. In a remote poverty-stricken village, he organized a two-week work camp where he and his students helped the villagers identify pressing needs and work together in meeting them. The process of acting together and recognizing their own resources of strength and intelligence generated strong feelings of community. This initial effort eventually led to the emergence of a movement based on people working together to meet common needs. Called Sarvodaya Shramadana, a Sanskrit phrase meaning "the awakening of all through working together," this movement spread across Sri Lanka to fifteen thousand villages.[9]

In its organizing strategy Sarvodaya applied the collaborative model of power with, working on the principle that everyone can play a role and that everyone has something to offer. At community kitchens set up to tackle childhood malnutrition, all those present, including the children, were encouraged from the start to contribute something, even if only some sticks for the fire or a matchbox of rice. In offering their time, ideas, and energy, people grow respect for their own abilities and deepen their sense of community.

In one village, a water reservoir had needed repairing for more than fifteen years. The villagers had built up a thick file of their correspondence requesting help from the local authorities. Then Sarvodaya organized a work camp of local and visiting volunteers to tackle the problem. They completed the job on the first day, and at a celebration held in the village that evening, they burned the file of letters.

Sarvodaya challenges the view that our society's problems are beyond the power of ordinary people to address. By acting together, we make things possible that before had seemed impossible. Working collaboratively toward a common benefit can also be deeply satisfying because it transforms work into a social occasion. This principle was used to great effect in the barn raisings common in North America in the eighteenth and nineteenth centuries. It is applied today in a wealth of community-building projects around the world.

The social fabric of Sri Lanka suffered decades of unraveling from civil war and ethnic violence, with tens of thousands of people killed between 1983 and 2009. During and after this time, Sarvodaya saw peace building as a process not just for high-level leadership to negotiate but also for everybody to play a role in. They organized "amity camps," inviting people from different sides of the conflict to work alongside one another in projects for community benefit. As well as building schools and clinics, these projects opened up a space for people to develop trust and friendship across divides and, through this, to participate in creating a more lasting peace.

The social unraveling that sets people against each other is, sadly, powerfully at work in our world. In 2020, hate crime reached its highest levels in the United States in over a decade, with more than ten thousand reported offenses related to race, gender, sexuality, religion, or disability.[10] An increase in hate crime internationally led UN Secretary-General António Guterres to speak out in May 2020, calling on governments and people to "act now to strengthen the immunity of our societies against the virus of hate."[11]

When looking beyond our bubbles of people in our home group, how welcome do we, and others, feel in the wider community around us? Like the second king in the story starting our chapter, do people in our neighborhoods sleep safely at night without fear of attack? Do we or others feel safe when walking down our nearby streets? The 2020 killing of George Floyd by a White police officer in the United States was an example of what the Black Lives Matter movement had been calling out for years: that Black people in the United States are at much higher risk than Whites of being killed by the police. If anyone feels less safe or valued in our neighborhood because of their race, gender, sexuality, religion, or disability, that diminishes the whole community.

The Great Turning begins with noticing the issues that concern us all. We move that story forward when we turn up with an intention to play our part. The question "What happens through us?" can provoke us to look for any ways our speech, thinking, or behavior might make others feel less safe, less equal, or less welcome. We can turn away from these actions, turning instead toward those that enrich social capital and strengthen community around us.

At a time of increasing polarization, another frontier for peace building is across the divide of political differences. Toxic polarization blocks the conversations and partnerships we need in order to address the problems we face. In the United States, Braver Angels is a grassroots organization that invites people to approach disagreement with curiosity and respect rather than hostility and avoidance.[12] Braverangels.org offers inspiring examples of people discovering a different kind of dialogue. Among those inspired are members of the US Congress who have called on the organization's founder to teach them its methods.

When viewed separately, initiatives like these might appear to have limited impact potential. Yet when we consider what they are part of and what they move toward, we can recognize their power. The Great Turning involves changing our culture. The way we interact and get along with the people in our wider communities is an essential element in this transformation.

THE GLOBAL COMMUNITY OF HUMANITY

Martin Luther King Jr. was arrested in 1963 for taking part in a nonviolent civil rights protest in Birmingham, Alabama. From his prison cell, he wrote a now-famous letter responding to criticism of the demonstration and the role of "outsiders" in it: "I am cognizant of the interrelatedness of all communities and states. I cannot sit idly by in Atlanta and not be concerned about what happens in Birmingham. Injustice anywhere is a threat to justice everywhere. We are caught in an inescapable network of mutuality, tied in a single garment of destiny. What affects one directly, affects all indirectly."[13]

"Outsider" involvement was criticized because of a belief that we should be concerned only with issues happening on our own doorstep. Dr. King dismantled this assumption. We don't need to live geographically close to people in order to care about them or to take action on their behalf. What extends community is solidarity based on the two Shambhala warrior tools: compassion and insight into our interconnectedness. Distance does dull us, though. If children were starving outside our front door, it would be unnatural to ignore them. Yet throughout our world, every ten minutes more than fifty children under the age of five die because they don't have enough food to eat.[14]

To bring closer the realities of our deeply divided world, consider what a village of a thousand people would look like if it reflected the composition of the world's population. Using data updated in 2019, the richest hundred people, one-tenth of the total, would have nearly 85 percent of the village's wealth, while the poorest hundred would struggle to get by on less than two dollars a day. About half the village would live on six dollars a day or less.[15]

In this scenario, 12 percent of the village would lack nearby access to safe drinking water, and nearly a third wouldn't have adequate sanitation. This would put them at risk of parasitic diseases, dysentery, cholera, and typhoid. During the first eighteen months of the Covid-19 pandemic, while most people faced enormous hardship, the richest ten people in the village would have seen their wealth increase dramatically.[16] While vaccines would have been easily available in richer parts of the village, with 50 percent of people fully vaccinated by September 2021, the rate would have been just 2 percent for those living in the poorer half.[17]

Looking into the future, the viability of the village is threatened by the pattern of overshoot and collapse noted in chapter 1. Because fresh water is being extracted at a rate faster than it is replenished, wells around the village are drying up. Because the land is overfarmed and topsoil is being lost to erosion, the area of productive agricultural land is shrinking. Because of overfishing, stocks of many once-common species have either collapsed or are in sharp decline. *Even without taking climate change into account*, it is easy to see that the village is heading for a crash.

If we lived in this village, would we see where we were headed? Would we pull together to address the challenges we face? Unfortunately, as reflected by our current global situation, the village would be divided into groups that see themselves as separate from and in competition with one another. A large share of the village wealth would be spent on military operations to keep resources in the hands of the richer groups, some of whom would be in conflict with each other. As these resources became depleted, wars would be waged over remaining reserves.

Thinking back to the two Danish kings, one might wonder whether sending young people to death and dismemberment in war is so very different from ordering them to jump from a tall fortified tower. In contrast, the Great Turning is about creating the sort of global community where people can sleep at night without fear for their safety.

When Helena Norberg-Hodge first visited Ladakh in 1975, one of the villagers told her, "We don't have any poor people here."[18] She saw he was speaking the truth: everyone's basic needs certainly appeared well met. There were no very rich people either — not in a material way, at least. In terms of social capital, though, the Ladakhi were among the wealthiest people she had ever known, and the happiest. They would sing together while bringing in the harvest. Their peacefulness was infectious. A phrase she kept hearing the villagers say was "We have to live together."[19] When a conflict arose, they would repeat this like a mantra and find a way of getting on. What would it be like if this were our mantra too?

We can choose between different types of wealth. The path of seeking material wealth beyond our basic needs sets us against one another. The greater a nation's appetite for resources, the more likely that it will go to war and the more likely that it will tear up forests for open strip mines or drill for oil deep below the ocean floor, wrecking marine habitats. Seeing with new eyes helps us grow another type of wealth. It is the community we find in mutual belonging.

The Earth Community of Life

Because she loved a river, in 2009 Ali Howard swam almost 380 miles in twenty-eight days.[20] The rich ecosystems of the Skeena in Canada were threatened by Shell's plans to drill a thousand gas wells around the headwaters of this river. Where small, scattered gas deposits were buried within seams of coal, high-pressure water and chemicals would be pumped into the ground as part of the extraction process. The contaminated silt that washed into local streams would threaten the spawning grounds of salmon not just in the Skeena but in the nearby Nass and Stikine Rivers too. To draw attention to the devastation these gas wells would bring, Howard swam the full length of the Skeena. Along her route, those living by the river came out to greet her, joining together in a newfound watershed identity.

Communities don't just involve humans; they include all that we belong to, feel part of, identify with, and act for. For Ali Howard, her community included the Skeena River itself and the rich ecology of plants, animals, and people within its watershed. When we stand up for a community, it is as though the community acts and speaks through us, making us its mouthpiece. Ali was one of many who gave the Skeena their voices to speak through. The grassroots coalition opposing Shell's plans included Tahltan and Iskut First Nations elders arrested for blockading the road to the river's headwaters, as well as First Nations councils, downstream communities, and conservation groups challenging the development. Together, they succeeded. On December 18, 2012, the government of British Columbia announced a permanent ban on drilling for petroleum and natural gas in a million acres of land that includes the headwaters of the Skeena, Stikine, and Nass Rivers.

This role, of speaking for our natural world, is crucial. If we don't, who will? Unless someone speaks for the salmon, the rivers, the wild spaces, and the rest of life, how will we stop the relentless drive of short-term profiteering that is turning our world into a wasteland? Our survival is at stake; we are only beginning to realize how ecosystems act together to maintain conditions favorable to humans. As James Lovelock, the leading scientist behind Gaia theory, explains, "The natural world outside our farms and cities is not there as decoration but serves to regulate the chemistry and climate of the Earth, and the ecosystems are the organs of Gaia that enable her to maintain our habitable planet."[21]

An understanding of our interdependence with all life is found in the wisdom of many indigenous cultures. As the Mohawk Thanksgiving Prayer states, "We have been given the duty to live in balance and harmony with each other and all living things."[22]

This duty is based on the recognition that without the interconnected web of ecosystems, we have no life. Yet we humans are living as if we were at war with the rest of nature. Between 1970 and 2016, wildlife experienced a catastrophic decline, with global populations of mammals, birds, fish, amphibians, and reptiles falling by 68 percent.[23] A major UN report released in 2019 warned that a million plant and animal species are now threatened with extinction.[24] We just don't know what impact the loss of these species will have, but activist and writer Duane Elgin offers a metaphor:

> Our extermination of other species has been compared to popping rivets out of the wings of an airplane in flight. How many rivets can the plane lose before it begins to fall apart

> catastrophically? How many species can our planet lose before we cross a critical threshold where the integrity of the web of life is so compromised that it begins to come apart, like an airplane that loses too many rivets and disintegrates?[25]

"We need to live together," the Ladakhi villagers say. "Or we will not live at all" is the message modern biologists would add. To stop the extinctions, we need to declare peace with our world. For that peace to take root and grow, we need active reconciliation and community building.

In the mid-1980s, Joanna, with John Seed, developed a group process that strengthens our felt relationship with other life-forms. Called the Council of All Beings, it invites us to step aside from our human identity and speak on behalf of another form of life.[26] It can be an animal, a plant, or a feature of the environment — an otter, an ant, a redwood, a mountain. We represent these life-forms at a gathering of beings who meet in council to report on the condition of our world. On one level we can see this as an improvised group drama, where we build empathy by looking through the eyes of another party. We could also approach this as a spiritual process, a ritual inviting a shift in consciousness that allows another part of our world to speak through us. Either way, we are dropping our normal lenses and taking a perspective that sensitizes us to the needs and rights of other beings.

In preparation, we take time "to be chosen" by the life-form we will represent. Then, in silence, we make masks. At the appointed time, often announced with the beat of a drum, we join together in a circle and listen as each being speaks in turn.

When we speak on behalf of another life-form, a shift happens in our relationship with it. Whether we've spoken for ants or for glaciers, bringing our imagination to bear in reporting their experience, they are no longer strangers to us. What emerges is a deepened appreciation of how they are affected by human activity and, with this, a sense of connection with their suffering and a desire that they be well.

Like Arthur drawing on the strengths of the creatures Merlin sent him to spend time with, we can experience the beings we speak for as sources of support. Here Chris describes a time when this happened for him:

> I was going through a difficult patch and sat down by a tree. I looked up and recognized the dark buds. It was Ash. I had been Ash and felt a sense of reunion with an old friend. Looking down, I saw Ivy. I had been Ivy too, at a different Council of All Beings. I felt supported by these two plants; I had a relationship with them I felt comforted by. My experiences of these councils have had a profound impact on my relationship with the life-forms I have represented. They have become significant as allies in my life. I wish to be an ally to them too.

This is the fourth dimension of community, in which we feel welcomed by our world and supported by it. Feeling part of a much larger team can anchor and steady us through times of difficulty. With this team spirit comes a heightened sense of spiritual connection with all life.

CHAPTER EIGHT

A Larger View of Time

In the largest fraud of its kind in history, Bernard Madoff swindled his victims out of more than $20 billion. For years, he maintained an illusion of profitability by siphoning capital from investments to pay his customers a return. In the long term, a Ponzi scheme like this is bound to collapse; the resource base of investments gets used up, and sooner or later there is not enough money left to pay people back. But until that point is reached and the fraud exposed, it seems a good bet.

While Madoff's actions were illegal, the practice of making quick bucks by plundering a resource that people depend on is not. Mainstream economics makes such short-term profiteering lucrative, even though it inevitably generates tragedies for the future. The problem is that costs are counted only if they appear in a very short window of time. The sorry tale of our oceans' fisheries provides a good example.

For hundreds of years, fishing communities along the coast of Newfoundland thrived, thanks to plentiful supplies of cod. From the 1960s onward, the use of larger ships equipped with sonar, onboard refrigeration, and much larger nets led to a massive increase in the catches. The cost of this, not counted at the time, was the almost complete elimination of the Northwest Atlantic cod population (see figure 12).[1] The same pattern of industrial-scale fishing has led to widespread collapse of fish stocks around the world. Once-common species, such as the Atlantic bluefin tuna and the common skate, are now threatened with extinction. If current trends continue, scientists predict that by the middle of this century commercial sea fishing could come to an end.[2]

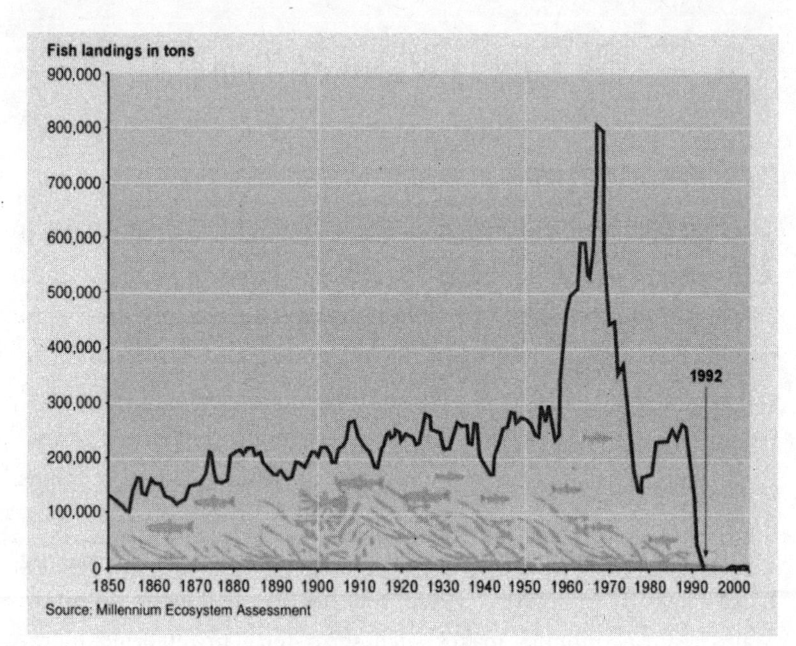

Figure 12. Growth in the fishing industry followed by overshoot and collapse.

Alan Greenspan, former chief of the US Federal Reserve, describes short-term thinking that ignores future costs as "underpricing risk," calling it a fundamental cause of the financial crisis of 2008.[3] His view is that our blindness to longer-term risks is a feature of human nature and, as a result, we're doomed to repeat the same mistakes. But is it human nature, or is it the product of a peculiar way of perceiving time that has, unfortunately, become mainstream in the industrial world?

When the Haudenosaunee meet in council to consider major decisions, their practice is to ask, "How will this affect the seventh generation?" This chapter describes how we too can live within a larger view of time. We will explore the concept of deep time and look at how it not only promotes greater ecological intelligence but also opens up new sources of strength, inspiration, and support.

A Family View of Time

When New College at Oxford University was founded in 1379, huge oak timbers were used to hold up the roof of its great dining hall. To provide replacement timbers for the roof's eventual repair, the college foresters planted a grove of oaks on college land. Since the beams were two feet wide and forty-five feet long, it would take several hundred years for the trees to grow to the size required. The foresters were thinking ahead in a time frame of centuries.

York Minster Cathedral in England took more than 250 years to build, while the temple complex of Angkor Wat in Cambodia was constructed over more than four centuries. The Neolithic structure at Stonehenge, with stones weighing four tons each that were transported from 240 miles away, is thought to have taken fifteen hundred years to complete. The craftspeople, architects, and planners of these projects must have accepted the fact that the work to which they were giving themselves would not be completed within their lifetimes or even within the lifetimes of their children.

Closer to home, a widened view of time is also found in the way most people think about their families. Families extend backward in time, and all over the world significance is given to lineage, ancestry, and family trees. When looking back at our family roots, at any point in history we will see a familiar constellation of parents, brothers, sisters, grandparents, aunts, uncles, cousins, children, and grandchildren. Over time, individual characters arrive, play their part, move through different roles, and then die; the larger system of the family, meanwhile, lives on.

When we see ourselves as part of a family system like this, we locate ourselves in a story that spans centuries. In each historical period people feel the same enduring motivation to care for the next generation, as the extended family acts through its members to keep itself alive. Hamish, a father of two young children, described his experience of this family mode of time by saying:

> I'm part of a family that is six foot tall,
> several yards wide,
> and hundreds of years long.

Families, of course, extend forward as well. In addition to our roots, shoots are still growing, and branches will sprout from them. If seeing ourselves as part of a family is an important aspect of our identity, then for how long would we like our family to continue? If the next generation matters to us, and the children born to it do as well, then what about their children and their children's children? Is there a point in time when we draw a line that says, "Beyond this they no longer count"?

This question may sound ridiculous, but unfortunately the line drawn in corporate and governmental decision-making comes frighteningly close. There is a time horizon beyond which consequences cease to be accounted for or even considered. Much of government planning works within a time frame of just a few years. When the long term is considered, the period looked at rarely stretches beyond a few decades. In major financial institutions, where the price difference between the morning and the afternoon can lead to profit or loss, even a few months into the future can seem farther ahead than is worth considering. Such extreme short-term thinking is a fairly recent phenomenon and is linked to an acceleration in our experience of time.

Time Is Speeding Up

In agricultural societies, the year's rhythm is counted in seasons. In the age before clocks, the sun moving across the sky gave shape to the day. Compare these natural cycles with the time intervals of modern technology, now measured in fractions of a microsecond. Life has become a race in a way that is historically unprecedented.

The sense of hurry is spurred on by an economic system that sets its goals and measures its success in terms of how fast it is growing. For an economy to grow from one year to the next, more needs to be accomplished in the same amount of time. If we want growth every year, then our speed of activity has to constantly increase.

Computer technology makes it possible to measure how fast companies are growing and to instantly compare vast numbers of them with one another. This has sped up the process of buying and selling company stock. In 1998 the average stock ownership period was two years. With the increased use of software-controlled automatic trading, stocks may now be held for only days, weeks, or months.[4] Much of this buying and selling is done by investment funds. Since the bonuses of fund managers are linked to how fast their funds are growing, they increasingly use their shareholder influence to push for maximum short-term returns. Company costs are cut through shedding staff, relying on casual or cheaper overseas labor, and neglecting maintenance. The constant pressure to improve returns leads employees to feel increasingly driven. Defending this approach, the managing partner of one of the world's leading private equity firms said: "A lot of public companies we speak to spent too much time on regulatory issues, social responsibility and corporate governance....And they forget their prime purpose — which is to grow the company as rapidly as possible."[5]

The Cost of Speed

The experience of speed can be pleasurable. There is a thrill to whizzing down a hill, an ease in instantly accessing information, and a joy in moving through tasks rapidly. When so much needs to be done to address our planetary crisis, speed is necessary. There is a big difference, however, between choosing to go faster when speed brings genuine benefit and being caught in a pattern of hurry through habit or demand. The value of speed is so embedded in our society, most of us end up feeling perpetually rushed and short of time.

Chronic hurry exacts a heavy toll. Straining against time affects our bodies with the release of adrenaline, tightening of muscles, and acceleration of the heart. While short bursts of pressure can be good for us, chronic stress wears us down, increasing our risk of heart disease, infections, depression, and many other conditions. Our relationships suffer too. A common factor in marital and family breakdown is a shortage of time to connect. In the long term, a sprint cannot be sustained, and burnout reduces performance. Crashing into a problem is often the wake-up call that alerts people to the hazards of high-speed living. If learned from, a crisis can become a turning point. The key to recovery, as we shall see, is a larger view of time.

When we are overwhelmed by numerous short-term goals and targets, we lack enough time and space to consider what lies over the horizon. Rushing narrows our field of vision to the immediate moment. The past becomes irrelevant and the future, abstract. Such narrow timescapes lead to the following five problems:

- Short-term benefits outweigh long-term costs.
- We don’t see disasters coming our way.
- Short-term timescapes are self-reinforcing.
- We export problems to the future.

- Short-term timescapes diminish the meaning and purpose of our lives.

Let us examine these problems in turn.

Short-Term Benefits Outweigh Long-Term Costs

The classic example of short-term benefits outweighing long-term costs is addictive behavior, where the allure of a cigarette, a drink, or a line of cocaine is its immediate effect. The downside of a behavior fails to deter us if it falls outside the bubble of time to which we give our attention. The same dynamic applies to dishonesty, which might seem a quick way out of a fix but has delayed effects that are toxic to relationships and decision-making. The issue of unsustainability arises from a similar narrowing of timescape. As the example of the fishing industry so clearly illustrates, the pursuit of continual growth in a resource-constrained world is a recipe for disaster.

We Don't See Disasters Coming Our Way

When we travel at great speed but look only a short distance into the future, disasters appear to come out of the blue. The story of the *Titanic* offers a grim lesson. Several days before the sinking, the captain had received warnings about icebergs over the radio. He had changed course slightly but not slowed down. On the evening of the crash, two ships had radioed warnings of icebergs, but the radio operator was snowed under with a backlog of personal messages and hadn't passed the warnings on. "Shut up, shut up, I'm busy," he'd replied when called at 11:00 p.m. by a nearby ship surrounded by ice. Forty minutes later, lookouts spotted a large iceberg straight in front. By then, the momentum of such a large ship traveling at close to full speed allowed no time for a change in course. Thirty-seven seconds later, the iceberg struck its fatal blow.

While there is always uncertainty when assessing future risks, some problems are reasonably foreseeable; they are like icebergs that our industrial society is racing toward. When we view these problems as far down the road in time, we treat them as nonurgent and leave them to one side. It is well-known that continued heavy consumption of fossil fuels is dangerously destabilizing our climate, yet there is great resistance to any slowing down of our oil-dependent economy. We would need the resources of another three to five planets like Earth if everyone lived the lifestyle of the average Westerner, yet the consumer lifestyle is still aggressively promoted throughout the world. Business as Usual can proceed only if we close our eyes to where it is taking us.

Short-Term Timescapes Are Self-Reinforcing

Living in a shrunken bubble of time is similar to being agoraphobic in that it is self-reinforcing. When looking into the future brings up despair and guilt, retreating into the "home" of a familiar timescape brings short-term relief. However, it also undermines motivation to act for our world, adding to the feeling that we are culpable. The way out of this trap is to recognize the important function guilt can serve and to honor this emotion as an expression of our pain for the world. Guilt is the uncomfortable awareness that our actions are out of step with our values. If we, collectively, don't experience guilt for what we're inflicting on future generations, we are in danger of pulling a Madoff on them. The beautiful world we were entrusted to protect will have been all used up.

We Export Problems to the Future

Our industrial economy is based on the practice of externalizing costs. This makes goods and services cheaper for those living now, though it leaves behind a debt incurred by us but not paid for. A common example is when a company reduces its safety and maintenance budget. This increases the risk of accidents, but because risk refers to events that haven't happened yet, they don't appear on the balance sheet. They are a cost that is projected forward in time.

When the *Exxon Valdez* oil tanker crashed in 1989, releasing oil that contaminated more than a thousand miles of Alaska's shoreline, the radar was switched off because it had been broken for a year and not repaired. Just ten months before this, oil company executives at a top-level meeting in Arizona had been warned that safety equipment was inadequate. The executives chose to save money rather than to address the problem.[6] Twenty-one years later, more than two hundred million gallons of oil poured into the Gulf of Mexico from BP's Macondo well. The presidential commission investigating this catastrophe found a similar pattern of cost cutting on safety, reporting that "many of the decisions that BP, Halliburton, and Transocean made that increased the risk of the Macondo blowout clearly saved those companies significant time (and money)."[7]

Fossil fuels and the things we use them for are artificially cheap because we don't count the costs we're passing on to future generations. Climate change is another example: with the world warming up, the ice sheets of Greenland and West Antarctica are already melting. The water these stores can release could raise sea levels by forty feet, which would cause flooding in two-thirds of the world's major cities.[8] While this might happen over an extended period of time, the images of New Orleans under water after Hurricane Katrina or of New York flooded after Hurricanes Sandy and Ida give a glimpse of what we are setting in motion.

Among our greatest crimes against future generations is the production of radioactive waste; through its mutagenic impact on the gene pool, its harmful effects are permanent. Because radioactivity is invisible, it will be difficult for future generations to know where the danger lies and be able to protect themselves. While the area that used to be Lake Karachay in the Chelyabinsk province of Russia might not look hazardous, it is so contaminated that a lethal dose of radiation can be received just by standing there for an hour.[9]

In the early 1950s, the lake was a depository for waste from the Mayak nuclear complex. Then, in 1957, at a nearby storage facility, the Kyshtym disaster occurred. A long-kept secret, this is one of the world's worst nuclear accidents. A fault in the cooling system of an underground storage tank led to an explosion that threw off the 160-ton concrete lid and blew 70 tons of highly radioactive fission products into the atmosphere. A dust cloud spread radioactive isotopes over 9,000 square miles, contaminating 270,000 Soviet citizens and their food supplies.

Since the 1950s, there have been more than a hundred serious nuclear incidents. While Three Mile Island, Chernobyl, and Fukushima are the best known of these, more than twenty others led to multiple fatalities and costs of over $100 million.[10] Each year, another twelve thousand tons of high-level waste are produced by the world's nuclear reactors. This waste contains isotopes that include iodine-129 (with a half-life of more than fifteen million years), plutonium-239 (with a half-life of twenty-four thousand years, but which decays into uranium-235, with a half-life of seven hundred million years), and neptunium-237 (with a half-life of more than two million years). We don't have a long-term solution for the safe disposal of nuclear waste. We don't know how to make containers that can withstand the embrittling effect of their radioactive contents. Much of the most toxic waste is kept in steel and concrete casks designed for a life span of a hundred years at best. We don't have a clear plan for what to do after this. It is a problem, and a cost, we are passing forward in time. As the fate of nuclear power stations built on earthquake fault lines at Fukushima in Japan so tragically illustrates, we are creating disasters that are waiting to happen, with harmful consequences that last virtually forever.

Short-Term Timescapes Diminish the Meaning and Purpose of Our Lives

Within the story of Business as Usual, a million years in the future, or even a thousand, is completely off the screen. When we are under time pressure, even ten years ahead might seem too distant to pay attention to. Dashing from one thing to another, we lose sight of where we're going. When life is dominated by urgent demands, we don't get the time needed to find our direction or determine what really matters. We can spend our days being busy without having our heart in what we do. This kind of busyness takes us further away from what's most important to us.

Thinking only in terms of abbreviated time periods also severely limits our sense of what can be achieved through us. To grow a project fruitful enough to be inspiring takes time. It is easy to ask, "What's the point?" if we're not seeing results after six months or a year. Imagine what would happen if we applied the same thinking after planting a tender young date palm or olive tree. These trees can take decades to become fully productive, but once they do they remain so for more than a century. When we move beyond thoughts of individual achievement and consider what our actions, if combined with the actions of others, can bring about, we open to a more gripping story.

Traveling in Time

In 1988 Joanna invited a dozen or so friends to join her in a study-action group on radioactive waste. During the previous years, she had asked scientists, engineers, and activists what they thought could be done with our nuclear facilities and materials once they were no longer in use. When she visited the construction site of a deep geological repository in New Mexico, the manager proudly told her about the high-tech barriers and signage that could deter intruders for a hundred years. "And after that?" Joanna had asked. He looked baffled. Larger time spans were outside his frame of reference, as they were for many others with whom she spoke. That is why she had invited the dozen or so friends. She wanted to create a space to explore human responsibilities grounded in a larger view of time.

The group took turns researching and presenting topics. One November day it was Joanna's turn, and the information she had gathered, about US nuclear waste containment practices, was both technical and horrific. To sustain attention and motivation, she needed some help. It came from an unusual direction.

On the door, Joanna put up a sign reading CHERNOBYL TIME LAB: 2088. As people came in, a recording of Russian liturgical music set the tone. Initiating their first group experience of traveling through time, she said:

> Welcome! Our work here at the time laboratory of this Guardian Site is based on the importance of being able to journey backward through time. This is because the decisions made by people in the late twentieth century on how to deal with the poison fire (or radioactivity, as they called it then) have such long-term effects. We must help them make the right decisions. So you have been selected to go back in time to a particular group in Berkeley, California, that has come to our attention. They are meeting exactly one hundred years ago today to try to understand, with their limited mentality, the ways their "experts" are containing the poison fire. It is easy for this group to feel ignorant and discouraged. Therefore, we will go backward in time to enter their bodies as they proceed with their study so that they will not become disheartened.
>
> Our research reveals that in time travel an essential factor is intention — strong, unwavering belief in the purposes the

heart has chosen. If our intention is clear, we can travel back a century to enter the hearts and minds of this very group. It will take about thirty seconds.

Joanna turned up the music for half a minute, then, switching it off, simply proceeded to present the day's topic. In the course of the session, no one commented on its odd beginning. Everyone was too keenly focused on the material itself. But there was a sense of heightened caring in the room, as though everyone sensed an added internal presence that was encouraging and supportive. By opening the group's imagination to the possibility that future beings can help us face the mess we are in, she helped the group discover a welcome source of additional inspiration.

The use of imaginary time travel, often sparked or accompanied by music, has since become a regular feature of the Work That Reconnects. It offers rich and rewarding opportunities to widen the timescape we inhabit and to draw on the support of past and future beings.

Ancestors as Allies

For much of the world's population, it is not so strange to think that those living elsewhere in time might be able to help us. Shrines to ancestors are common in Japan and Korea, and the practice of seeking guidance from those who have lived before us is an accepted part of many indigenous traditions. As West African shaman and author Malidoma Somé writes: "In many non-Western cultures, the ancestors have an intimate and absolutely vital connection with the world of the living. They are always available to guide, to teach and to nurture."[11]

The interest ancestors take in us is a natural extension of the care parents have for their children or grandparents for their grandchildren. When we're struggling or feeling alone, we can find moral strength in opening to a sense of ancestral support. Just as an athlete may perform better when cheered by a crowd, we can imagine a crowd of ancestors cheering us on in all that we do to ensure the flow of life continues.

If the role of the ancestors is to look out for those who come after them in time, then it follows that we too play this role. Those living in the future will look back on us as their ancestors. Recognizing future beings as our kin brings them closer to us. Within the abbreviated timescape of Business as Usual, they are a forgotten people whose interests have disappeared from view. When we recognize that we are their ancestors, a sense of care and responsibility arises naturally.

Connectedness with ancestors and with future generations lifts us out of the microplots of Business as Usual and places us in a truer and more expansive story. In life's epic journey, every one of our ancestors lived long enough to pass on the spark of life. This ancestry extends back in time far beyond the reaches of our human past. With the shift in identity to our ecological self, we discover that the entire span of recorded history is just a fraction of a page in a more extensive volume.

Our Journey as Life on Earth

We belong to a planet four and a half billion years old. To make relative time periods easier to grasp, let us look at the entire history of our Earth as if it were a single twenty-four-hour day starting at midnight.[12] In this day of planet time, each minute would mark the passing of more than three million years (see figure 13).

At first, the planet was as hot as an erupting volcano. Being formed by the gravitational pulling together of materials orbiting the sun, it was continually showered by meteorites. Shortly after midnight, a chunk of matter the size of a small planet collided with Earth, the impact causing materials to be thrown outward into space to form the moon. It took till nearly two in the morning for the planet surface to cool down enough for steam in the atmosphere to condense into rainfall. As rain fell and kept falling, the oceans were born.

Between three and four in the morning, the first forms of life appeared in warm, shallow water. There were only traces of oxygen in the atmosphere and no ozone layer to provide protection from incoming ultraviolet radiation, which was too strong to allow life to develop on land. Photosynthesis took until ten-thirty in the morning to evolve, and from then on, those early green life-forms started producing oxygen as a waste product.

All life-forms were single celled and remained so for the rest of the day, the first more complex multicelled organisms not evolving till half past six in the evening. By eight, worms had appeared at the bottom of shallow seas, followed an hour and twenty minutes later by the first fish. By a quarter to ten, plant life had become established on land; soon after ten, amphibians and then insects appeared.

At twenty minutes to eleven, disaster struck in what has been described as the mother of all mass extinction events. A combination of volcanic activity, asteroid impacts, and other catastrophes wiped out 95 percent of life, though that left plenty of room for the dinosaurs to emerge afterward as the dominant vertebrates on land. The age of the dinosaurs lasted till twenty to midnight, when a six-mile-wide meteorite struck Earth and caused a dust cloud to block out so much sunlight that the decline in plant life killed off many large animals. Mammals, who had been quietly in the background for the past hour, emerged to fill the niche of dominant vertebrates on land. Ten minutes later, some mammals returned to the sea and slowly evolved into whales and dolphins.

At two minutes to midnight, a small ape in Africa became the last common ancestor of both humans and chimpanzees. At just twenty seconds to midnight, apelike hominids discovered the use of fire. The entire history of our species, from its early origins in Africa, is contained in the last five seconds before midnight.

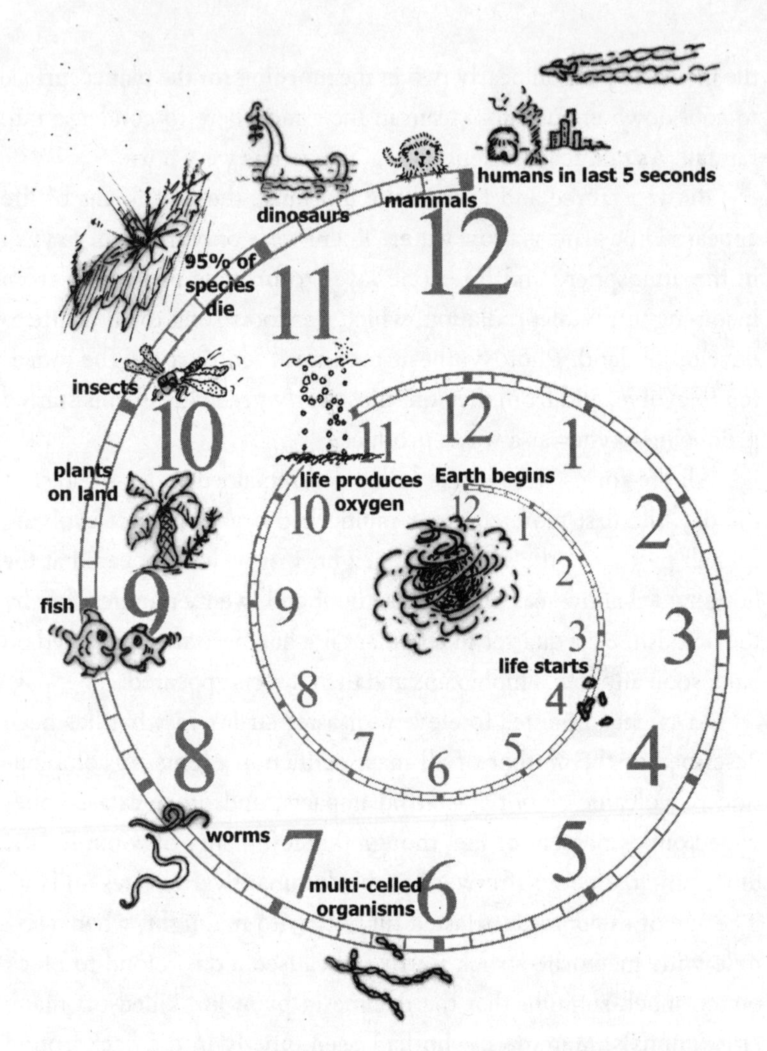

Figure 13. Four and a half billion years of our planetary history is represented as a single day. Each hour represents 187.5 million years. Each minute represents 3.125 million years.

Within the story of Business as Usual, a commonplace saying is "You can't change human nature." But when we look at the breathtaking span of our planetary history, the idea that we'll never change seems absurd. We are part of the most extraordinary unfolding. Where will it go next?

Our Journey as a Species

To get a clearer sense of the critical point in human history we have reached, we can take that last five seconds of the planet's day and represent it as a twenty-four-hour time period (see figure 14). We are moving from planet time to human time, where all our history is represented by a single day.[13]

While apelike relatives using stone tools and fire had been around a million years ago, our species, *Homo sapiens*, is thought to be a quarter of a million years old. If we take the past 240,000 years as our day, again starting at midnight, and represent each 10,000 years by an hour, we didn't leave Africa till six in the evening. For 95 percent of our history, we lived in small groups as hunter-gatherers (as some indigenous people still do today), turning to agriculture only at ten to eleven in the evening. A few minutes later, one of the first recorded cities, Jericho, arose.

By twenty past eleven at night, we'd discovered the wheel and were beginning to develop early forms of writing. By half past eleven, work on Stonehenge had begun, and the first city-states were developing in what is now Egypt, China, Peru, Iran, the Indus Valley, and the Aegean Sea area. At a quarter to midnight, the Buddha and Confucius were both alive, with Jesus coming a few minutes later and Muhammad a few minutes after that.

At five minutes to midnight, windmills were in use, and two minutes later, Christopher Columbus reached what we now call North America. In this full day representing human history, the industrial age didn't begin till two minutes before midnight. In the last minute, the world population rose from just over a billion to more than seven billion. In the last twenty seconds (that is, since

1950), we have used up more resources and fuel than in all human history before this.

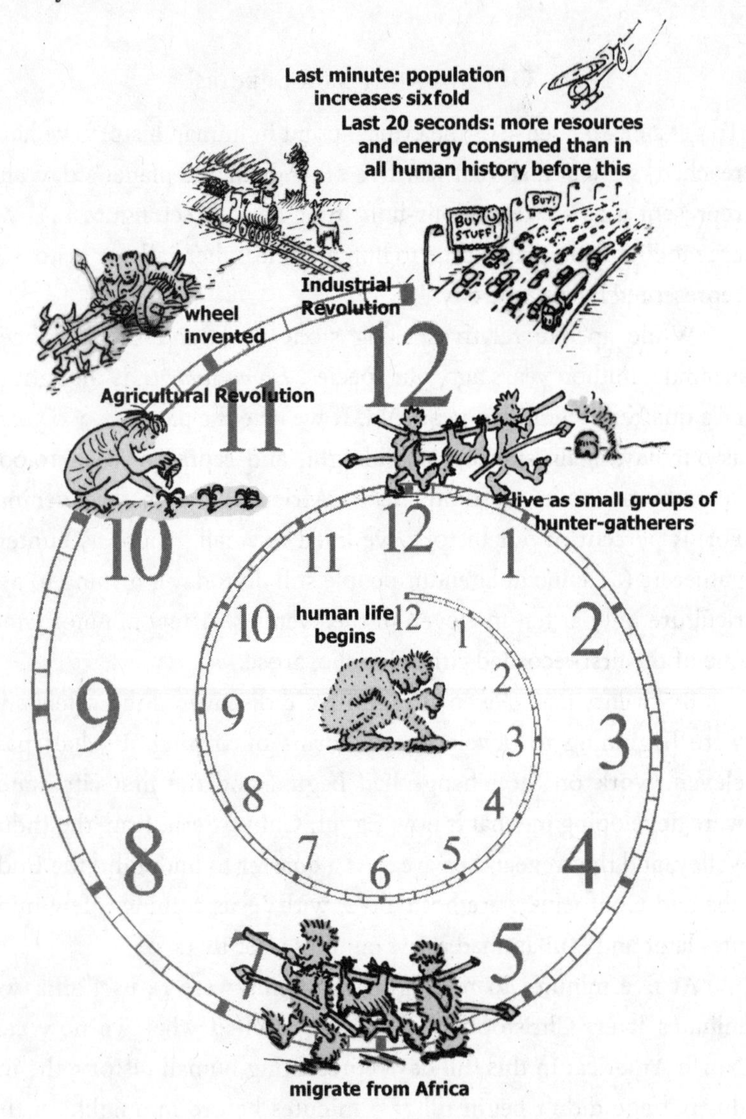

Figure 14. Here, 240,000 years of human history are represented as a single day. Each hour represents 10,000 years; each minute, 166 years.

The last third of a minute of this day of human history has seen an explosion not just of our population but also of our appetite for energy and resources. In the Business as Usual culture of industrialized economies, it is considered normal to consume thirty-two times the quantity of resources and to produce thirty-two times as much waste as those living in nonindustrial societies.[14] What is not being seen is how much harm we are causing. Because our culture is not looking at where it is going, we are like the *Titanic*: heading very quickly for a crash.

Learning to Reinhabit Time

A wonderful advantage of widening our time span is that doing so opens up a sense of possibility. If land-based mammals can return to the oceans and evolve into dolphins, then it is not such a stretch to think that modern human beings could return to a state of connectedness with the land and evolve into a wiser form of life.

The development of our species has been marked by — and driven by — discoveries that increase our capacity: speech, writing, tools, agriculture, the wheel, and now communication technologies that enable us to cooperate with people we've never met living thousands of miles away. Could the discovery, or rediscovery, that ushers in the ecological age be the ability to reinhabit time?

Ecological intelligence involves thinking in terms of *deep time* — a temporal context that includes our whole story. We need to do this now because, given our technologies, our actions have consequences extending millions, even billions, of years. Take the thousand tons of depleted uranium weaponry used in Iraq and Afghanistan. The cancer-causing aerosol it leaves behind has a half-life of 4.5 billion years. That is as long as the age of Earth.

Learning to live in a larger timescape opens us to new allies and sources of strength. Our ancestors can be our allies, and we ourselves, as the ancestors of future generations, can play the role of ally to them as well. Perhaps these future generations have something to say to us.

Many of the advances we take for granted today were thought of as impossible before they were invented. Things we consider impossible today might still be developed in the future. Could future generations, for example, discover a way to communicate with us? And if so, what might they say? Perhaps they could do this only if we played our part too, by extending ourselves forward in time to meet them. We can do this through our imagination. We don't know whether the communications we would receive this way are real or imagined — and we don't really need to know. They still offer useful guidance. Here is an exercise we often use in workshops, which we value for the richness it brings.

TRY THIS: LETTER FROM THE SEVENTH GENERATION

The purpose of this exercise is to help us identify with future beings we might think of as the "seventh generation," that is, those living about two hundred years from now; to see our efforts from their perspective; and to receive counsel and courage.

Closing your eyes, imagine journeying forward through time to identify with a human living two hundred years from now. You do not need to determine this person's circumstances, only to imagine that they are looking back at you from the time they inhabit. Imagine what this person would want to say to you. Open your mind and listen. Now begin putting words down on paper as if this future one were writing a letter just to you that starts: "Dear Ancestor..."

By giving future beings a voice, we bring them closer in a way that helps us be guided by their perspective. Hearing ourselves reply to them also helps us to step into a larger experience of time. This next exercise invites us to reply to questions asked by future beings.[15]

Try This: Letter to the Future

Imagine that the Great Turning took place in the twenty-first century, allowing human life to continue, and that future beings will know about what happened in our time. Now write a reply to three questions you have received from those future ones:

1. "Ancestor, I hear stories about the period you are living in, with wars and preparations for war, with some people absurdly rich while huge numbers are starving and homeless, with poisons in the seas and soil and air, and with the dying of many species. We are still experiencing the effects of all that. How much of this do you know about? And what is it like for you to live with this knowledge?"
2. "Ancestor, we have songs and stories that tell of what you and your friends did back then for the Great Turning, but they don't tell us how you got started. You must have felt lonely and confused at times, especially at the beginning. What were some of the first steps you took?"
3. "Ancestor, I know you didn't stop with those first actions on behalf of life on Earth. Where did you find the strength to continue working so hard, despite all the obstacles and discouragement?"

We can bring deep time to mind as we go about our daily lives. Even as we wash the dishes, pay the bills, or go to meetings, we can school ourselves to be aware, now and then, of the hosts of ancestral and future beings surrounding us like a cloud of witnesses. We can remember the vaster story of our planet and let it imbue the most ordinary acts with meaning and purpose. Each of us is an intrinsic part of that story, like a cell in a larger organism. And in this story, each of us has a role to play. As we move into the going forth stage of the spiral, we begin to focus on that role.

PART THREE

Going Forth

CHAPTER NINE

Catching an Inspiring Vision

On August 28, 1963, Martin Luther King Jr. delivered what is now one of the best-known speeches in history. More than half a century later, the words "I have a dream…" are still linked to the vision he shared that day.[1] King described a future in which Black and White children would join hands as brothers and sisters and in which his own children would be judged by "the content of their character" rather than the color of their skin. He was identifying a destination that could be reached, a reality that could be created. In the 1960s the idea that an African American might one day become president of the United States would have been dismissed as "just dreaming." While the Black Lives Matter movement reminds us how far there is still to go, just look at the impact that kind of dreaming can have.

Our dreams and visions for the future are essential for navigating through life because they give us a direction to move in. As the Roman philosopher Seneca once remarked, "If one does not know to which port one is sailing, no wind is favorable." Moreover, moving toward a destination that excites and inspires us energizes our journey, puts wind in our sails, and strengthens our determination to overcome obstacles. The ability to catch an inspiring vision is therefore key to staying motivated. When we're moved by a vision that we share with others, we become part of a community with a common purpose.

Inspiration is often thought of as a fleeting experience that strikes us in a lucky moment or as a rare strength found in the small number of people thought of as "visionaries." In this chapter we look at how we can foster vision and inspiration using learnable skills. The insights and practices discussed here can help all of us become more inspired and visionary. They strengthen our capacity for going forth in the adventure of the Great Turning.

How Our Imagination Gets Switched Off

From an early age, we are schooled in a worldview that values facts over fantasies. The term *dreamer* is used as a dismissive put-down when someone's ideas are considered unrealistic; daydreaming in the classroom can even be a punishable offense. To develop our visioning ability, let's start by recognizing how it has become so undervalued. As Quaker futurist Elise Boulding comments: "Several generations of children have had daydreaming bred out of them. Our literacy is confined to numbers and words. There is no image literacy."[2]

After decades of research into how the brain works, we're beginning to understand that our two cerebral hemispheres function in different ways.[3] While the left side thinks in terms of words and rational logic, the right side works more with images and patterns, helping us to integrate complex information and to sense the larger shape of things. Since our educational system focuses almost exclusively on words and numbers, it is as though we are being taught to use only half our brain. A simple first step in developing our visioning ability is to recognize it as a form of intelligence that is both valuable and learnable.

To underline the crucial importance of visioning, consider how many aspects of our present reality started out as someone's dream. There was a time when much of what is now the United States was a British colony, when women didn't have the vote, and when the slave trade was seen as essential to the economy. To change something, we need to first hold in our mind and heart the possibility that it could be different. Stephen Covey, in his bestselling book *The Seven Habits of Highly Effective People*, writes: "'Begin with the end in mind' is based on the principle that all things are created twice. There's a mental or first creation, and physical or second creation to all things."[4]

Imagining possible futures is a surefire way to develop foresight. If we're interested only in "facts," we limit ourselves to looking at what has already happened, which is a bit like trying to drive a car by looking only in the rearview mirror. To avoid crashing, we need to look where we're going. Since we can't know for sure what will happen, we are limited to considering possibilities by applying a combination of experience, awareness of trends, and imagination. While experience equips us well for dealing with familiar situations, our imagination is essential in formulating creative responses to new challenges.

Liberating Our Imagination

A design principle that boosts creative thinking is "What comes before how." First identify *what* you'd like to have happen; working out *how* comes later. If we exclude options just because we can't immediately see how they can happen, we block out many of the more exciting possibilities that might inspire us. An important distinction is to be made between the creative phase, in which we generate ideas and possibilities, and the editing phase, in which we choose and evaluate them. Placing an embargo on editing in this first stage liberates our creativity.

Our intention in the creative phase is to catch a vision so compelling that it touches us emotionally. To remain motivated during difficult times, we need to *really* want our vision to happen. When what we hope for seems beyond our power, however, we are likely to hear a voice inside us saying, "There's no point in even thinking about this; it just isn't going to happen." To hold on to an inspiring vision, we need to stop ourselves from shooting it down inside our minds before it has a chance to take form. What helps here is recognizing the difference between static and process thinking.

Static thinking assumes that reality is fixed and solid, resistant to change. When people say things like "The problem is human nature; it is never going to change" or "You can't change the system," they're taking this approach. They are viewing situations as if they are like pictures hanging on a wall: if a new idea or way of doing things isn't already in the picture, they see it as unrealistic (see figure 15). This view limits our sense of what is possible; if nothing inspiring is on the horizon, we can easily fall into apathy and resignation.

Figure 15. Static thinking sees reality as fixed and resistant to change.

With *process thinking*, we view reality more as a flow in which everything is continually moving from one state to another. Each moment, like a frame in a movie, is slightly different from the one before. These tiny changes from frame to frame generate the larger changes seen over time (see figure 16). If something is not in the picture at the moment, that doesn't mean it won't be later on. This way of conceiving reality sees existence as an evolving story rather than as predefined. Because we can never know for sure how the future will turn out, it makes more sense to focus on what we'd like to have happen and then to do our bit to make it more likely. That's what Active Hope is all about.

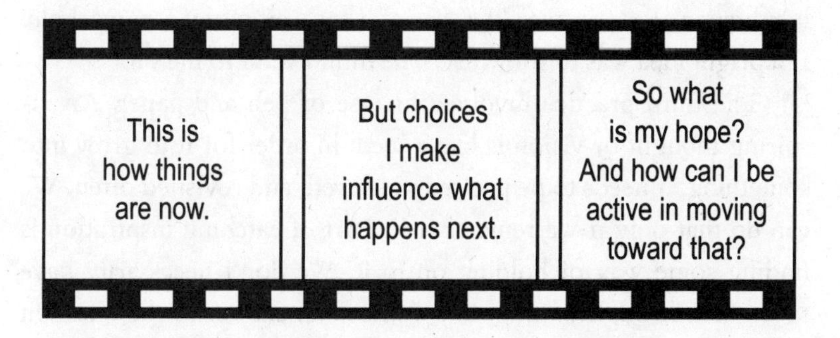

Figure 16. Process thinking: each moment is like a frame in a movie.

Practices That Help Us Catch Inspiration

There was once an inventor who sat for hours every day in a special soundproofed room. With a pencil in his hand and a pad of paper on his desk, he was waiting for ideas. When they came, he would make a note and then return to his waiting. This inventor is modeling three practices that can help us catch an inspiring vision.

The first is simply creating space. When we're too busy, our attention is so occupied that it leaves little room for anything new to enter. That is why people often remark that their most inspiring moments come when they are on vacation, out for a walk, or taking a shower. Allowing ourselves quiet moments, even to daydream, opens a space into which inspiring thoughts can flow. Such pauses can be surprisingly productive: Thomas Edison came up with many of his inventions while lying on a couch in his workshop. It was a daydream about a snake eating its tail that led German chemist

Friedrich August Kekulé to discover the ring structure of benzene molecules.

The second practice for catching inspiration brings together two tools freely available to us all: intention and attention. By placing himself in the soundproofed room, the inventor was making himself fully available to any creative impulses that might arise. His heightened alertness was like that of a cat waiting by a mousehole. If a bright idea was to show itself, he didn't want to miss it.

The third practice involved his use of pen and paper. An inspiring thought or vision is like a seed: in order for it to grow into something, it needs to be planted, nurtured, and revisited often. We can do that only if we remember it. Part of catching inspiration is finding some way of holding on to it. We don't necessarily have to do that through writing, but we need an answer to the question "How will I remember this a year from now?"

These three practices are the key ingredients of a guidance system that helps us find and follow the purposes that call us. There are many different ways of making space, of focusing our intention to catch an inspiring vision, and of anchoring what comes up so that we don't forget it. We'll be looking at a few approaches in more detail; it is worth experimenting with a range of them to find the ones that suit you best. The core principle here is that we can do more than passively wait for inspiration — we can invite it in. We can also train ourselves to tune in to visionary signals.

Since the best way of anchoring a vision is to act on it and make it part of our lives, we need a way of linking our larger hopes with specific steps we can take. Visioning therefore involves three closely related stages:

1. **What?** When looking at a specific situation, what would you like to see happen?
2. **How?** How do you see this coming about? The second stage involves describing the steps needed in order for

the larger vision to occur and possible pathways by which these steps could take place.

3. **My role?** The first level identifies the desired destination, the second level maps out the story of getting there, and the third identifies your role in this story: What can you do to help the vision come about?

Imaging the Future We Hope For

Observing that the peace movement was clearer about what it was against than what it was for, Quaker futurist Elise Boulding developed a workshop called Imaging a World without Weapons. Its structured process moves through the three levels of visioning we've outlined, with a sequence that maps out more detailed steps. Here is how she describes the beginning of the process:

> There are two basic components to this type of imaging. The first is intentionality. We must allow ourselves not only to wish for but to intend the good. Our intentions about the future must be serious, so the first exercise in the imaging workshop is to make a list of things we want to see happen in the world thirty years into the future. We choose thirty years because it is far enough into the future so something can have happened, and near enough so that many of us will live into that time. Participants must make their own lists. The items on it become their intentions.
>
> The second component of imaging involves unlocking the total store of imagery in our mental warehouses.... To help participants get into that fantasying mode, the second exercise is to step into a childhood memory. Participants are asked to pick a happy one. I often go back to age nine and climb an apple tree in our yard. It is essential to use all one's inward senses in experiencing the scene; see the colors, feel

> the textures, smell the smells, hear the sounds, note the expressions on people's faces. . . . After a couple of minutes of remembering, each participant turns to a neighbor and describes where he or she has been, then listens to where the neighbor has been. Sharing the experience helps to clarify the memory.
>
> It is this type of imaging you do when you image the world thirty years from now. You will see a world without weapons around you with the same vividness, as fragments from your stored experiences get recombined in a kind of movie reel inside your head.[5]

As preparation for moving into our preferred version of the future, the third stage of the workshop invites us to imagine ourselves standing in front of a huge, thick hedge stretching as far as we can see to the left and right. On the other side of this lies the world we hope for thirty years from now, where the intentions we identified in the first stage have been realized. We can move through or over the hedge any way we want and have a good look around. The goal is to be an observer and to collect information that can be reported back on.

After the imaging, we are invited to share with others what we have experienced. The more specific our observations are, the more real the image becomes to the person we're describing it to. So rather than using generalizations such as "No one is hungry," we are encouraged to be detailed in describing what we saw or heard. Each of us sees only a fragment of this potential future, so the next and fourth part of the workshop involves piecing together information yielded by different people's imaging to build a more coherent picture of what this future offers. How are decisions made? How does education work? The imaging provides clues that groups can build on. Boulding continues: "We are in no way prophesying or predicting or compelling the future by this kind of imaging, any more than we are compelling the future when we pray. The enactment of the future depends on what we do, how we individually and collectively respond to what we envision."[6]

The workshop has two more parts. After spending time working together on developing a vision of our preferred future, we need to imagine how we got there. Standing in the world thirty years from now, we look back at how the changes we have just envisioned would have been brought about. Moving back year by year, what happened? As we trace back to the present moment, we construct a "history" of the next thirty years from the perspective of this possible future.

Finally, we need to see the role we play in this process. Looking at the different areas of our lives, what are we doing that helps build the future we hope for? There will be many things, and some might stand out — what are they?

Imaginary Hindsight

A powerful mental shift takes place when we stop telling ourselves why something can't happen. When we can envision a hoped-for future, we strengthen our belief that it is possible. By inhabiting this vision with all our senses, imagining what colors and shapes we see, the expressions on people's faces, the sounds we hear, the smell, taste, and feel of this future, we bring ourselves there in a way that activates our creative, visionary, and intuitive faculties. As the poet Rumi once wrote: "Close both eyes to see with the other eye."[7]

Research has shown that when people approach a problem by imagining that it has already been solved and then look back from this imagined future, they are more creative and detailed in describing potential solutions.[8] Permaculturalist and activist Rob Hopkins used this imaginary hindsight approach when founding the Transition movement. He told us: "The idea didn't emerge fully formed all at once. It was more a case of thinking, 'I wonder what would happen if...' and then imagining what a permaculture response to the challenge of climate change would look like, particularly a response that could easily catch on around the world."[9]

The Transition movement idea that each community could develop its own "energy descent plan" grew out of this visioning process. The starting point is an image of what a community would look like if it were no longer dependent on oil and other fossil fuels. How would it function? What would people eat? How would its economy, healthcare, and education systems work? Then a trail is traced back in time from that possible future, marking key developments along the way. The pathway sketched out offers the beginnings of an energy descent plan by which the community can wean itself off fossil fuel dependence. Finally, as in Boulding's model, we end in the present, with our lives here and now, looking forward and identifying how we can play our part in the transition process.

Imaginary hindsight can be applied to a range of timescales. If we're facing a challenge in the next twenty-four hours, we can imagine its successful resolution a day from now and then look back from that point at what we did to get there. The story of how we rose to the challenge directs our attention to the steps we need to take.

THE STORYTELLERS' CONVENTION

It can be fun as well as illuminating to play with this imaginary hindsight process through the use of storytelling. When giving talks, Chris sometimes guides audiences in imaginary time travel to a hoped-for future hundreds of years from now. He invites them into the fantasy that they are at a gathering of storyteller historians where, in pairs, they take turns sharing their stories of the Great Turning. Taking just a few minutes each, they tell the tales they might have heard as children about that historic period in the twenty-first century when human society seemed on the path to collective suicide. Though things didn't look too promising at first, a widespread awakening occurred, and huge numbers of people rose to the challenge of creating the life-sustaining society familiar to the storytellers now. After sharing their stories, the audience travels back to the present time, carrying with them a deepened sense of the mythic adventure they are part of.

While the stories told are fantasies, some may be close to realities that will occur. When we carry our desired future inside us, it guides and acts through us, helping us bring it into being.

Nightmares Can Inspire Us Too

We have been using the word *dream* to refer to the futures we hope for. However, it is not only positive visions that guide us. Our nightmares can alert us to dangerous conditions and summon us to respond on behalf of life. In the fall of 1978, not long after offering her first group experience of "despair work," Joanna had a particularly lucid and disturbing dream. Engaged at the time in a citizens' lawsuit to stop faulty storage of high-level radioactive waste, she had been reviewing public health statistics in localities around nuclear power plants. One night, before going to bed, she had leafed through baby pictures of her three children to find a snapshot for her daughter's high school yearbook. That night she had a nightmare:

> I behold the three of them as they appeared in the old photos. Their father and I are journeying with them across an unfamiliar landscape. The terrain becomes dreary, treeless, and strewn with rocks. Little Peggy can barely clamber over the boulders in the path. Just as the going is getting very difficult, even frightening, I suddenly realize that by some thoughtless but unalterable prearrangement, their father and I must leave them. I can see the grimness of the way that lies ahead of them, bleak as a Martian landscape and with a flesh-burning

> sickness in the air. I am maddened with sorrow that my children must face this ordeal without me. I kiss them each and tell them we will meet again, but I know no place to name where we will meet. Perhaps another planet, I say. Innocent of terror, they try to reassure me, ready to be off. Removed and from a height in the sky, I watch them go — three small figures trudging across that angry wasteland, holding one another by the hand and not stopping to look back. In spite of the widening distance, I see with a surrealist's precision the ulcerating of their flesh. I see how the skin bubbles and curls back to expose raw tissue, as they doggedly go forward, the boys helping their little sister across the rocks.

When Joanna described this dream in an article, the dread of a mother seeing her children left behind in a barren, toxic world struck a strong chord with readers, many of them writing to share similar nightmares and their own fears for our world.

The Dream of the Earth

When we think of ourselves as interconnected parts of a larger web of life, just as we may feel the Earth crying within us, perhaps we can experience the Earth dreaming within us too. Within the framework of systems thinking and the ecological self, this makes perfect sense. If we think of self as a stream of consciousness with many currents, then inspired dreams and visionary moments are simply times of increased connectedness with the deeper flows of our collective identity.

The idea that our dreams (those we have when awake or asleep) can tap into deeper sources of wisdom is found in many cultures. In the Bible, the Pharaoh's dream of seven lean cows eating seven fat cows was interpreted by Joseph as a warning that seven years of famine would follow an equal period of abundance. For the

Sonenekuiñaji, an indigenous community in the Peruvian Amazon, dreams are valued as essential guides that inform their decisions in daily life.[10] In *The Dream of the Earth*, Thomas Berry writes, "Such dream experiences are so universal and so important in the psychic life of the individual and of the community that techniques of dreaming are taught in some societies."[11]

The simplest technique is just to pay attention to our dreams, to see them as significant. We might make a note of them while they are still fresh in our mind. Some cultures have early-morning gatherings to share their dreams of the night before and draw out possible meanings. Our dreams can tell us things, but only if we listen to them.

If we're able to receive guiding signals from our larger ecological self, then, rather than catching inspired visions, perhaps we could say it is the inspiring vision that catches us. In every situation, latent possibilities are waiting to come into being. In dreamy visionary states we get glimpses of these possibilities; when these visionary images land in us, they can work through us and take form in our actions. The same vision can catch several people and bring them together in a community of shared purpose. In this view, a visionary impulse is not something we invent; it is something we serve.

While this concept of guiding visionary signals fits comfortably in both spiritual and systems perspectives, it clashes with the ultra-individualistic worldview of the industrialized world. In this model, thinking and intelligence are located within individuals; if someone comes up with a good idea, that idea is regarded as their property. When the ownership of ideas is privatized, innovations that could help our world are often kept secret until they can be exploited for personal or corporate benefit. What would happen to a brain if separate groups of neurons took this approach?

Thinking happens through brain cells rather than within them, and intelligence is an emergent property of cells working together as a larger whole. Viewing ourselves as similar to brain cells opens us up to an entirely new way of thinking about intelligence. Co-intelligence, which we touched on in chapter 6, is defined by the Co-Intelligence Institute as "accessing the wisdom of the whole on behalf of the whole."[12] If we thought of ourselves as part of the team of life on Earth, could we access the wisdom of our world for the benefit of all?

Co-intelligence arises when we share ideas and visions we find inspiring, making room to hear what moves other people too. This is how visions catch on and spread through a culture with amazing speed.

Try This: Sharing Inspiration

The following open sentence can be used as a prompt for a round in groups, as an opening to conversation with a small group of friends, as a starting point for personal journaling, or as something to include in letters to friends:

- Something that inspires me at the moment is...

Co-intelligence in Progressive Brainstorming

A progressive brainstorm process that Joanna has used in workshops provides a good example of co-intelligence arising. This process is a great way of harnessing group creativity in moving from a larger general goal to specific steps that participants can take. It can be fun and energizing to recruit a few interested friends to try this process together. The first stage is to identify a goal to work toward. Each person daydreams for a few minutes about what he or she would like to see in a life-sustaining society. They then list these features on a big sheet of paper. Here's an example.

Stage 1: Our Vision for a Life-Sustaining Society

- Clean air
- People's essential needs met, e.g., food, shelter, water
- Renewable energy
- Constructive processes to deal with conflict
- Lifestyles of voluntary simplicity

For the next stage, the group selects one of these items, asks the question "What would be needed for this to happen?," and has a brainstorm session to generate responses. As the goal of brainstorming is to spark creative thinking, the process is guided by three rules. First, we don't censor, explain, or justify our ideas. Second, we don't evaluate or criticize the ideas of others. And third, we save discussion for later. We're creating options, not editing them.

Once the group has generated a list, they choose one of the options and repeat the brainstorming process, once again exploring the question "What would be needed for this to happen?" Each time this is repeated, the steps identified become closer to us, in terms of feeling more within our range of influence and therefore easier to take action on.

Stage 2: What Would Be Needed for Us to Have Clean Air?

- No incinerators
- Scrubbers on smokestacks
- More renewable energy investment
- Walking or cycling replacing car use on short trips
- Shift to electric vehicles

The same brainstorming process can then be applied to each of the new line items. The goal is to end up with a list of practical steps that anyone in the room can take. When this process is in full swing, you can feel the group thinking and strategizing through each of its members.

Stage 3: What Would Be Needed for Walking or Cycling to Replace Car Use?

- Bicycle repair classes
- Community walking events
- Higher fuel prices
- More public transportation
- Bicycle delivery services

The process of selecting and brainstorming continues until the group lands on an idea so specific and appetizing, it can be dramatized in a role play.

CHOOSING AND BEING CHOSEN

With every theme we take up, a range of possible pathways will arise for actions in which we could play a role. With so many options, how do we choose where to invest our energy? The challenge is to listen for the vision that calls us most strongly and to recognize that to follow this well, we will need to refine our focus so as not to dissipate our energy. Like seedlings that need thinning out, we need to choose which visions we support and then clear space around them so that they have room to develop and thrive.

Sometimes we experience such a strong intuitive pull toward a course of action that we know it to be the right thing for us to do. Even when the odds seem against us, we feel these powerful summonings deep in our hearts, and we are drawn to respond.

In the model of co-intelligence, we're never alone in these endeavors. A larger story is taking place, and we've just chosen — or been chosen — to play a particular role in it. If we trust in a larger intelligence, we can open to the support of many allies and helpers who will play their roles too. Joseph Campbell wrote, "Follow your bliss... and doors will open where there were no doors before."[13]

We don't make these visions happen — we just play our part in them. To do that, we need to keep our vision and our commitment to it strong inside us; then we can follow wherever it takes us. Here's an example from Joanna's life.

Journey to Tibet

The vision took hold when a message arrived from a dear friend and teacher she had known for more than twenty years. A Tibetan lama in exile from the Chinese occupation of his homeland, Drugu Choegyal Rinpoche was based in Northwest India with his refugee community of monks and laypeople. His friendship with Joanna began in the 1960s, when he was eighteen and Joanna, twice his age, was living in India with her family as part of the American Peace Corps. Over the years, through Choegyal Rinpoche's stories and luminous paintings, the Tibet he had left behind became very real to Joanna.

He never expected to see it again, but in 1981 and for the next half dozen years, a moderation in Chinese occupation policy allowed lamas in exile to visit. Choegyal Rinpoche's trip to his homeland in the province of Kham in eastern Tibet was so rewarding to him and so inspiring to his people that he proceeded to plan another visit in 1987. This time he asked Joanna, now living in California, to join him there.

> I wrote back an immediate YES! and said that my husband Fran and daughter Peggy would be coming too. Choegyal

Rinpoche's paintings of his homeland were so vivid to me that I could picture being right there in the high green hills of Kham, and just as compelling were the plans we had been talking about. Choegyal was eager to restore his monastic center from the ruins left by the Cultural Revolution, gathering some of the monks who had gone into hiding; and for the laypeople he would build a Tibetan-language school and a craft center for traditional arts. Having been inspired by my stories of the Sarvodaya self-help movement in Sri Lanka, he called his vision the Eastern Tibet Self-Help Project.

I was fifty-eight years old then, and our daughter would turn twenty-six on the trip. The only Tibetan map I found that featured the region of Kham unfolded awkwardly to a good three feet wide. On it, Kham, about the size of Switzerland, appeared no bigger than the palm of my hand. It showed the route we planned to take, approaching Tibet from the east through Szechuan. Taking over the dining room in our home, Peggy and I assembled oxygen tablets, water filters, rolls of film, vitamins, Band-Aids, belly medicines, trail mix, high-altitude lip gels, sunglasses, and gifts for our Tibetan hosts. We each bought lightweight mountain boots — in beige and navy for Peg and Fran, aqua for me.

We secured our Chinese visas but not the papers allowing us to enter the Tibet Autonomous Region, which until recently had been closed off. The consular agent in San Francisco assured us that we could obtain these Alien Travel Permits after we landed in Beijing. We banked on this hope as the three of us checked our bulging backpacks and sleeping bags and settled into our seats on the Air China plane.

At each successive stop — Beijing, Chengdu, and the smaller towns farther west, where we pressed on for days by bus and hired car — we beseeched every Chinese authority we could find for the permits to travel within the heartland of

Tibet. Every single one refused. Our last hope was Derge, the town closest to the Yangtze River, which served as the border itself. But there the police not only refused us the permits, they also ordered us to leave the entire area within two days.

The determination to join Choegyal in his homeland was as strong as ever. It was even intensified by the efforts we had already expended and by the tantalizing proximity to Kham in the Autonomous Region of Tibet, barely fifty miles away on the other side of the Yangtze. So on hearing that a high lama I knew, long based in the West, was passing through Derge on his way out of Tibet, I sought his help. The Venerable Akong Rinpoche greeted us warmly, but he had no hope to offer, though he had good relations with Chinese authorities. Indeed, two border officials, coming from their duties at the Yangtze checkpoint, dropped by for tea while we were there. When Akong Rinpoche sounded them out on our behalf, their answer was curt and final.

Someone else, however, was in the room, quietly listening. Tenzin was a young Tibetan from Derge, about to return to his studies in Berlin. He overheard me say the words "I will do anything to get in. Anything." Tenzin repeated them back to me when he tracked us down that night. "If that is so, you must love your lama very much. I have a plan for you, if you are willing to try it." The plan involved a local man he knew and trusted, a man who had a tractor. We met this man within the hour, when Tenzin brought him to meet us and look at our map. Since we spoke no Tibetan, we used gestures only, and his were few. It was best not to know his real name, so Fran, Peggy, and I, among ourselves, called him Tractor.

Early the next morning, we met Tractor in the marketplace; he barely glanced at us. Impassively, he nodded toward the narrow wagon hitched to his three-wheeled tractor. Eight others, including a couple of monks, had climbed aboard and now pulled us up to sit beside them on the narrow boards along each side. Soon we were slowly rumbling westward out of Derge, along fields of barley stubble shining golden under a brilliant blue sky. Our wagon mates' smiles and chatter made it seem as if we were all heading off on a picnic.

By the time we stopped on a high shoulder above the Yangtze three hours later, all but one of our traveling companions had left us, replaced by heavy bags of barley pulled aboard at recent stops. Peering downriver, we saw the far end of the bridge, sentry boxes, figures moving. Quickly pulling us out of view, Tractor gestured us onto the wagon's narrow floor and hushed us into silence. No sooner did we lie down — we could only manage it on our sides, spoon fashion, Fran curved around my back, me around Peggy's — than the sacks of barley landed on top of us, pinning us down in airless darkness.

With Tractor's companion sitting on top of us, the weight was crushing, but the struggle to breathe was the hardest thing. As we jolted down the road headfirst, my chest heaved and hammered for air. I prayed to my lungs: Be still. I prayed to pass out before exploding in panic. I prayed that Peggy not suffocate. After a turn and a couple of stops, shrill Chinese voices were shouting just a foot or two away. The wagon started up again, made another turn, finally stopped. Sheer gratitude eased the breathing even before the sacks were pulled off us, but then we saw Tractor's face. In pantomime he conveyed what had happened: the rifles aimed at his head, the sacks at the end of the wagon pulled aside to expose three pairs of American boots.

The wonder was that we hadn't been hauled out on the spot and tossed into jail. For that, we figured, we could probably thank those border officials who took tea with Akong Rinpoche and knew that three American friends of his were attempting to cross the border. Another wonder was that Tractor did not abandon us forthwith to protect his own life. Instead he doggedly persisted in trying to get us across the Yangtze. By late afternoon he had brought us some nine or ten miles downriver from the bridge, turning off the one hillside road down a steep dirt track toward the water's edge. Switching off the motor, he walked us with our gear to a grove of trees where a rowboat lay hidden. As daylight dimmed, he ferried us across.

The skiff was shallow, the current strong. It took two scary trips to get our bodies and packs to the farther shore, all the while in full view of Szechuan's one paved road above the Yangtze. On the rocks below the steep, wooded bank, we turned to say good-bye to Tractor. He was coatless now and shivering. He knew, as we did, that we had to move fast if we were to find our way north to where the road came off the bridge and then head farther into Kham before dawn. He pointed to the river, the sky. Yes, we understood, keep the river on our right, follow the stars. I touched his arm, trying uselessly to thank him, trying to memorize the face of this man whose name I didn't know. I wanted to be able to recognize that face someday, on a street, in a crowd.

Four weeks later Fran, Peg, and I are seated once more aboard an Air China plane, this time on our way home. In my lap lie notebooks filled with jottings, figures, lists, sketches. Rough estimates of cement and wood to be procured are mixed in with drawings by Choegyal Rinpoche of the school he envisages and dimensions of the looms he proposes for the craft center. Names of the families who'll volunteer labor figure alongside those who have agreed to contribute rocks for the walls or food for the building crews. I know all this will serve me well in writing the grant proposal for the modest funds we need. And I know that the Eastern Tibet Self-Help Project has already begun.

Equally precious to me are the faces that drift before me, as the plane's lights dim for the night. In my mind's eye I see Tractor and Tenzin and the others who helped us in essential ways we could not have foreseen. I see the Khampa horseman whose campfire we found soon after our river crossing. With his teenage son and a second horse, he brought us through the moonless night on a narrow trail cut into cliffs above the Yangtze. I see the tiny old woman from the village we reached by sunrise. No one else there would notice or speak to us, causing us to conclude that near the border the local Tibetans were more fearful of the Chinese. Yet after we had walked on out of sight and waited long, vain hours for a lift, she appeared, hobbling slowly down the road. In a brass kettle as heavy as she was, she was bringing us hot tea. I see the truck driver who took the risk to break the rules and take on three foreigners who'd been marooned for days in the mud, shit, and grease of a transport yard. And, of course, I see my old friend Choegyal Rinpoche standing motionless by his glistening white tent at the heart of his people's vast mountainside encampment, greeting us with calm pleasure, as if he'd had no doubt for a minute that we would make our way to him.

CHAPTER TEN

Daring to Believe It Is Possible

In 1785 Thomas Clarkson, a student at Cambridge University, entered an essay competition; the topic was slavery. As he researched the subject, he was horrified by what he discovered about the transatlantic slave trade. His essay won first prize, but its content so disturbed him he found it difficult to sleep. In his diary he wrote: "In the daytime I was uneasy. In the night I had little rest. I sometimes never closed my eye-lids for grief."[1]

Abandoning his plans to become an Anglican minister, he decided instead to join a small group of Quakers campaigning on the issue. Two years later, Clarkson was one of the dozen people whose meeting in a London printshop kicked off the Society for Effecting the Abolition of the Slave Trade.

Along with William Wilberforce and a few other committed campaigners, Clarkson had caught an inspiring vision. At the time, the odds seemed completely stacked against them. Slavery was accepted as a normal part of life. The slave trade was well established, and powerful vested interests opposed any change. In the British House of Commons, the idea of abolishing the slave trade was described as "unnecessary, visionary and impracticable"; repeated attempts by Wilberforce to bring in a new law were blocked. Such was the opposition that Clarkson was nearly killed when he was attacked by a mob on the docks of Liverpool.

When following the path of an inspiring vision, we are likely to encounter the voice dismissing what we hope for as unnecessary or impractical. The greater the gap between present reality and what we would like to have happen, the louder this voice will be. Yet stories like those of Clarkson and Wilberforce remind us that when we dare to believe our vision is possible and dare also to act on this belief, extraordinary changes can take place. In 1807 the British Parliament passed a law making the slave trade illegal within the British Empire. As other countries followed, slavery became outlawed throughout much of the world within little more than a lifetime. A quote widely attributed to Nelson Mandela describes changes like this: "It always seems impossible until it is done."[2]

Recalibrating Our Hopes to Start from Where We Are

While daring to believe in an unlikely vision can at times help breakthroughs happen, there are also occasions when it can be reckless or even harmful to do so. For example, how realistic is it to believe that we can continue growing our industrial economies when we're already experiencing such harmful consequences and there's worse on the horizon? Yet this is the vision Business as Usual is based on.

How can we tell the difference between visionary hopes that inspire great change and those that are no longer viable or that take us in a harmful direction? Helpful here is the three-step framework of Active Hope. We begin from where we are, taking in a clear view of reality. Then, from that starting point of awareness, we identify what we hope for. The third step is to head in that direction. That first step is essential, because it invites a recalibration of hopes so that they are based on where you are now, taking account of any changes in circumstances and new information.

When we consider the range of ways a situation can go, an image that supports both realistic appraisal and openness to possibility is the double spider diagram (see figure 17).[3] This begins with a drawing of a spider on its side, with its legs fanned out to one side. The spider's body represents the present moment, and each leg is a different version of how things might develop. Upper legs are better options, lower legs are worse.

If we've followed a lower leg and things have worked out badly, we can apply the spider view again from a second point in time. Thus we get the double spider: the first presenting the range of possibilities under normal conditions; then, acknowledging a turning for the worse, the second inviting us to consider different versions of how the next bit of the story might go.

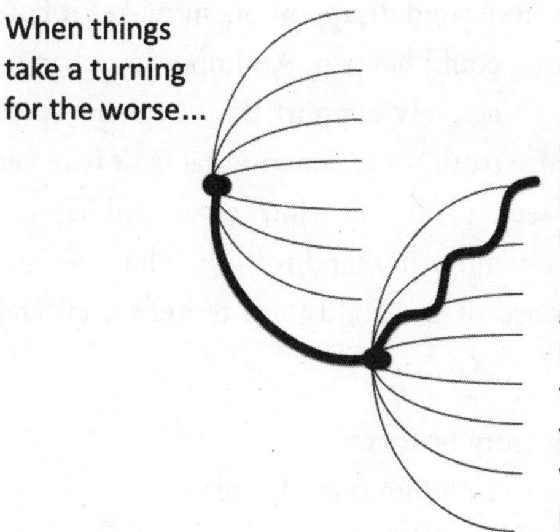

Figure 17. The double spider diagram shows a range of ways a situation might develop.

Over recent decades, the facts of climate change, habitat loss, mass extinction of species, and extreme inequality have become more and more discouraging. Some of the hopes that seemed achievable even a decade ago are now more difficult to sustain. Letting go of a hope, relinquishing it, can be a positive act when it is a feature of acknowledging the new conditions we face. Yet at the same time, considering the best that can happen starting from where we are nourishes the conviction that our actions might make a difference.

Acceptance and commitment therapy (also known as ACT) is a well-established treatment for depression and anxiety.[4] The twinned qualities of acceptance and commitment are like the two spiders. The first involves acknowledging where we are, recalibrating hopes, and relinquishing those no longer supportable. The second spider view lets us look at what is still to come, choosing what to get behind and give ourselves to. This is about commitment.

It is a judgment call as to when to let go of a vision and when to believe one is still possible. Repeated disappointment makes it hard to even consider the best that could happen. An important change-making skill is the ability to actively support the visionary space that can see a desirable future truth — a scenario that isn't true yet, and which might not even seem likely, but which could still happen, particularly if we deliberately move in that direction. The reference points that support our sense of possibility are therefore crucial. Let's look at five of these:

- Inspiring examples from history
- The phenomenon of discontinuous change
- Facing your threshold guardians
- Our own experiences of perseverance
- Witnessing the Great Turning happening through us

Inspiring Examples from History

It took Clarkson and Wilberforce more than twenty years of campaigning before they saw a law against the slave trade passed.[5] They were challenging the Business as Usual mode of their day, and while they had periods of hard-earned popular support, they also experienced times when their cause looked completely hopeless. In the early 1790s, the French Revolution and Great Britain's subsequent war with France led to a clampdown on political opposition in Britain. Harsh new laws forbade gatherings of more than fifty people unless they had permission from a local magistrate. If a magistrate declared a meeting illegal and more than twelve people were still gathered an hour later, they could be sentenced to death. The abolition committee gave up its London office and went for seven years without meeting. Clarkson had a nervous breakdown, and the campaign went into decline.

There is a saying that important changes often go through three phases. First they are regarded as a joke. Then they are treated as a threat. Finally, they become accepted as normal. Historical examples of changes moving through this sequence provide an important reference point for us. If it is a struggle for you to believe that what you hope for is possible, know that others have felt this way too. Thanks to the people who kept alive the spark of their convictions through periods of ridicule and persecution, changes that were laughed at, actively suppressed, or dismissed as hopeless dreams have now become accepted as normal parts of our reality. Here are some examples:

- Most people now accept that the Earth orbits around the sun.
- Women have the vote in nearly every country in the world.
- Apartheid came to an end in South Africa.
- An African American was elected US president.

Galileo was branded a heretic for believing the Earth revolved around the sun, saw his writings banned, and spent the last eight years of his life under house arrest. Lucy Stone organized the first National Women's Rights Convention in Worcester, Massachusetts, in 1850. She did not live to see women finally get the vote in the

United States in 1920, but that didn't stop her from working her whole life to make that happen. Nelson Mandela was imprisoned for more than twenty-five years before eventually becoming South Africa's first Black president in 1994. When we experience frustration, failure, or obstruction, we need to remember that we are part of a long and fine historical tradition.

Could it be that one day future generations will look back and wonder how we dared to believe we could create a life-sustaining society — and dared also to bring it about? For that to happen, we need the capacity to tolerate frustration. Rather than seeing frustration and failure as evidence that we're pursuing a hopeless cause, we can reframe them as natural, even necessary, features in the journey of social change.

Failure and frustration may well be necessary parts of the journey, because if we stick only with what we know how to do and feel confident about, we limit ourselves to the old, familiar ways rather than developing new capacities. Learning is a process that begins with not knowing. When we learn a new skill, we tend at first to get it wrong, gaining experience through our mishaps. The good news about frustration and failure is they show that we have dared to step outside our comfort zones and rise to a challenge that stretches us. Let's reframe frustration and failure in a way that encourages us to persist rather than give up. By persisting, we are more likely to experience the positive surprise of reaching the next reference point.

THE PHENOMENON OF DISCONTINUOUS CHANGE

With an issue like climate change, we don't have seventy years, or even twenty years, to bring about the changes needed. Our problems are urgent and demand much more rapid action. Given our current slow pace of progress, it can be hard to see how we will do it.

If we look at change as something that happens incrementally at a steady, predictable rate, in which the progress — or lack thereof — in one decade gives a measure of what is likely to take place in the next, we can get discouraged. But along with continuous change, there is also *discontinuous change*. Sudden shifts can happen in ways that surprise us; structures that appear as fixed and solid as the Berlin Wall can collapse or be dismantled in a very short time. An understanding of discontinuous change opens up a genuine sense of possibility.

Consider what happens to a bottle of water when it is left in the freezer. As it cools down, its temperature undergoes a steady, continuous change. The water won't change much in appearance until it begins to get near the critical threshold of its freezing point. Then, as it passes this, an extraordinary process happens. Tiny crystals form, and when they do, other crystals form around them, building a mass movement of crystallization in the water that rapidly changes state from liquid to solid. This is discontinuous change.

With discontinuous change, a threshold is crossed where, rather than just more of the same happening, something different occurs. There's a jump to a new level, an opening to a new set of possibilities. We might think it impossible that a small amount of water could crack something as hard as glass, but as the ice expands, it breaks the bottle.

Even when we don't see a visible result from our actions, we may be adding to an unseen change that moves the situation closer to a threshold where something crystallizes. When a line is crossed and something new starts to happen, the change may appear to come out of nowhere.

Discontinuous changes can be triggered by quite small events. When you're close to a threshold, one tiny step can take you over it. An example of this is when a critical mass of people starts to believe a change can happen, creating a tipping point. If those who are undecided have been hovering on the edge beforehand, waiting to see what happens, something small can tip the balance and nudge them into giving their support. Before this threshold is crossed, the change may seem unlikely. Yet only a short while later, everyone wants to join in.

Each time we take a step forward, we don't know how this action will interact with the actions of others to create an entirely new set of circumstances. When one change makes other changes more likely, it can trigger a cascade of events that leads to a breakthrough. Like a snowball rolling downhill, each movement forward adds momentum. With synergy, unexpected shifts can occur as if by magic.

Discontinuous shifts can happen in both directions. Just as new ways of thinking can catch people's imaginations and cause pivotal innovations to take off, so too can tipping points bring about rapidly worsening conditions. While we can be poised on the edge of a major evolutionary leap, we can also face the very real danger of catastrophic collapse.

We can't know how things will unfold. What we can do is make a choice about what we'd like to have happen and then put ourselves fully behind that possibility. An inspiring wave of change is spreading through our world. The Great Turning is happening in our time, and we may already be participating in many ways. If we want this change to catch on more thoroughly, deeply, and rapidly, how can we let it do so in our own lives? What thresholds do we need to cross?

Facing Your Threshold Guardians

When we catch our inspiring vision, we experience a call to adventure that pulls us into its service. There may be a treasured piece of forest we want to protect, a community project we want to support, or a valuable quality we wish to bring out in ourselves and others. But we always seem to bump into some kind of resistance or opposition. Mythologist Joseph Campbell coined the term *threshold guardian* to describe whatever is guarding or blocking the way.[6] Studying myths, legends, and adventure tales from all over the world, he mapped out a common plot structure that revolved around the tension between the impetus to follow the call to adventure and the threshold guardians standing in the way.

In fantasy adventures such as *The Lord of the Rings*, whenever there is a clear path to follow, some kind of monster or other enemy is bound to show up. The hero or heroine then rises to the challenge by confronting, tricking, befriending, or bypassing the blocking entity in a way that allows their journey to continue. We can view our own lives similarly. When we're following our calls to adventure, thinking of obstacles as threshold guardians can draw out our creative response. Sarah, an activist with the Transition movement, described the impact doing this had on her:

> I wanted to set up a Transition project in the city where I live. But I thought, "No way, it is too huge a project, I can't do that." Hearing the idea that our lives are like adventure stories changed the whole way I thought about this. It was a huge thing to do, and I wasn't sure it was possible, but that's the way these stories begin.

Whenever Sarah was struggling, she'd say to herself, "This is my adventure; these are my threshold guardians." Rather than feeling defeated, she knew she needed to seek out allies, learn new skills, and not be put off when things weren't working out.

Humans have been telling adventure tales for thousands of years, not just for their entertainment value but also because they pass on teachings that help us rise to challenges. Almost always, the protagonists find their way through by drawing on the fellowship of allies, by serving a purpose much bigger than themselves, by discovering strengths previously hidden from view, and by having enough humility to learn from others. Usually some mysterious force lends a hand. It could be a spiritual presence, the moral force of a value like justice, or, as in the film *Avatar*, the emergent power of an interconnected web of life.

Sometimes the threshold guardians have a tangible physical presence outside of us. For Joanna, following her powerful summoning to meet Choegyal Rinpoche in Tibet, the blocking force came in the form of the visa-refusing authorities and the Chinese border guards. But the thresholds we cross lie as much inside us as they do outside. For Sarah to step into her commitment to set up a Transition City initiative, she needed to find a way past the voice that said, "No way, it's too huge, I can't do that." This is the threshold guardian of disbelief.

It can help to imagine these blocking voices as characters standing in our way. Some years ago, Chris worked with a group of puppeteers to give visible expression to the common blocks of fear, cynicism, and disbelief. Fear took the form of an overprotective parent constantly warning of the dangers of stepping into anything new. The voice of cynicism was a dismissive distant relative who tore apart the value of any project considered. Finally, disbelief was personified by Professor Noway, a very clever character who has studied everything and knows precisely why there is no way we can succeed.

These characters sometimes have useful things to say. Listening to our fear can stop us from being too reckless, and a healthy skepticism protects us from getting involved in causes that don't ring true. Professor Noway might, in fact, be right from time to time. When encountering the inner voice that would stop us, how do we know whether it is protecting us or blocking us? Just asking this question can help us consider both possibilities. By listening to our own objections and challenging them if they don't stand up to scrutiny, we can counter our own resistance. When we undo the blocks that hold us back, we liberate our power to act. The next reference point illustrates this.

Our Own Experiences of Perseverance

Many of us will have struggled to make a change that seemed so difficult we wondered whether it was even possible. The times when we have eventually found a way through offer another important reference point of possibility. They remind us that when we hear the voice saying, "No way, I can't see it happening," we can sometimes prove Professor Noway wrong. Here Chris recounts an example from his own life:

> In the late 1980s, I worked as an intern/junior hospital doctor in London. As was common practice then, my contract was for an average of eighty-eight hours a week, though some weeks I would work well over a hundred hours. Starting work at eight in the morning, I might not stop work till three in the morning the next day. But as I was still on call, I could be called back to work half an hour later, eventually getting to bed at five in the morning for a brief snatch of sleep that might last only an hour before I was called back to the ward again. I would then continue throughout the day till my work finished at five that afternoon.
>
> I would work around the clock like this for two thirty-three-hour shifts most weeks, plus working during the day on the other weekdays. On top of this, every third weekend I would have a continuous stretch of duty of between fifty and eighty hours at a time. During some of these weekend shifts (which could extend from Friday morning till Monday afternoon), the longest period of uninterrupted sleep I had at any one time was ninety minutes.
>
> When I started working like this, I found the conditions

difficult to believe. I was surprised there wasn't a mass campaign to improve conditions, but nearly all the doctors I spoke to felt resigned and powerless. "We don't have the power here," I was told. "There is nothing we can do." As I asked around, nearly everyone agreed that although it was a crazy way to work, there was no point in trying to change things. First, any campaign would be doomed to fail because the system was so deeply entrenched; second, if you spoke out, you'd be threatening your career, because doctors were dependent on good references from their employers in order to get their next job. If you wanted to move forward in medicine, you had to put up with the hours and keep your mouth shut.

A small group of doctors didn't let these obstacles stop them and had set up a campaign. I immediately joined them. Together, we wrote letters to newspapers, organized lobbies of Parliament, and arranged meetings in hospitals. As the campaign began to grow, a television news team approached me for an interview. The evening before it was due to take place, I was phoned at home by a hospital manager. "If you go ahead with this," I was told, "you'll need to watch out for your future." The message was clear: speaking out was career suicide. Yet this was the fear that kept the system in place. If there was ever to be a change, it would require that people stick their necks out. So I went ahead.

Moving across that threshold of fear was liberating, and as other doctors started to speak out too, a small committed group set up a media campaign. When we carried out an overnight protest outside a London hospital, the issue became a major news story. Over the next year, we did more high-profile actions and received lots of press coverage. Unfortunately, there was no change in our working conditions, and morale in the campaign group began to dip.

Suffering from severe sleep deprivation, I became clinically depressed. A friend who'd been a lawyer before studying medicine suggested legal action. Together we consulted a well-known barrister. In his opinion, there was no basis in law for a claim, and attempting legal action could damage the campaign, since it might get laughed out of court. A second lawyer came to a similar verdict: "There's no precedent," he said.

Another friend, also a lawyer, took a different view. She offered to help, free of charge, in attempting to secure an injunction to stop the hospital requiring me to work more than seventy-two hours in a week or more than twenty-four consecutive hours without a protected period for sleep. Every employer had a legal obligation to provide a safe system of work; the health authority was failing in this, because exhausted doctors were more likely to make mistakes, and sleep deprivation was a risk factor for depression. To maximize media coverage, the writ was served on a bank holiday, a Monday just after I had finished my fifty-hour weekend shift.

Thirty journalists and six television crews were waiting outside the hospital. "Yawn this way," they called as they took photos. The story made international headlines, and friends on the other side of the world saw the news on their televisions. As a media stunt, it worked brilliantly. But drawing attention to the issue wasn't enough to bring about the change that was needed.

At a court hearing, the claim was judged to be a frivolous action with no basis in law and was struck out. That could have been the end of it. But a leading barrister stepped forward to offer his services for free. He believed there were grounds for an appeal, though if it were lost, I would be charged for the health authority's legal costs, and I didn't have the funds to cover them. A newspaper article described our dilemma, and people started sending in money. More than two hundred letters of support came flooding in, along with thousands of pounds in donations.

The doctors' union, the British Medical Association, was approached to see if they'd back the case, but their legal adviser said it didn't have a chance. In a newspaper interview, Lord Denning, the country's top judge, gave the same opinion. It would be unheard of, he said, for a court to declare that working too hard was illegal. Yet when I read the letters of support, including some from doctors who, like me, had felt so desperate they'd considered suicide, I knew I had to continue.

Taking a holiday in Scotland, one afternoon I momentarily fell asleep while driving. When I opened my eyes, I was on the wrong side of the road with a car speeding toward me. I swerved out of the way, and my car went out of control, veered off the road, and crashed into a rock face. The car was totaled. Fortunately I escaped with just a graze to my knee. But the message seemed clear: if I continued working like this, driving myself on and on without adequate sleep, I was going to crash. Just a few days previously, I'd finished a working week of 112 hours.

The crash served as a wake-up call. The working conditions weren't just unpleasant; they had become a matter of life and death. Around that time, several young doctors had been killed in car accidents, and sleep deprivation was suspected as a cause. I gave notice of my resignation and became more determined than ever to pursue the case. A few months later, the appeal succeeded. The ruling here was an important victory: it set a legal precedent establishing that it was unlawful for an employer to demand so much overtime as to bring foreseeable harm to health. At this point, the British Medical Association decided to back the case, and preparations for the trial began.

Over the next few years, the health authority made several more attempts to block the case on legal grounds. They even tried to take it to the House of Lords. But eventually a trial date was set for May 1995. It was to be a category A trial, meaning it was judged to be of great public significance. Just weeks before the trial was due to begin, the health authority offered an out-of-court settlement. They were concerned they might lose and wanted to avoid the costs of an expensive trial. They paid me a small damages award, plus all the legal costs. Along with the precedent set by the Court of Appeal ruling, this settlement sent a powerful message to employers: they would be financially vulnerable to negligence claims if they failed to consider the health impact of excessive hours of work. The age of hundred-hour workweeks was coming to an end. The case was won.

When a change wants to happen, it looks for people to act through. But how do we know when a change wants to happen? We feel the want inside us. We sense a desire, a tugging at us to be involved. But that doesn't make the change inevitable, because standing in our way are all those who say we're wasting our time, that it isn't possible, that it will be too hazardous. For the change to happen through us, we need to counter those voices. A shift can happen within us when we break through a resistance that has been holding us back.

Witnessing the Great Turning Happening through Us

When we see with new eyes, we recognize how every action has significance, how the bigger story of the Great Turning is made up of countless smaller stories of communities, campaigns, and personal actions. We don't know whether these changes will happen fast enough to prevent a catastrophic collapse, but given the unpredictable nature of discontinuous change, they just might. This is not the same as being confident, but it does involve being open to the possibility of success.

If you were freed from fear and doubt, what would you choose to do for the Great Turning? Here is an action planning process we use in our workshops. It will help you to identify some practical steps you can commit to taking in the next seven days. Seeing is believing. When you see yourself take these steps, it is easier to believe that the Great Turning is happening.

TRY THIS: IDENTIFYING YOUR GOALS AND RESOURCES

The following process works well when you team up with someone else, taking turns to interview and support each other. You can repeat and review this process regularly.

1. If you knew you could not fail, what would you most want to do for the healing of our world?
2. What specific goal or project could you realistically aim to achieve in the next twelve months that would contribute to this?
3. What resources, inner and outer, do you have that will help you? Inner resources include specific strengths, qualities, and experience, as well as the knowledge and skills you've acquired. External resources include relationships, contacts, and networks you can draw on, as well as material resources such as money, equipment, and places to work or recharge.
4. What resources, inner and external, will you need to

acquire? What might you need to learn, develop, or obtain?

5. How might you stop yourself? What obstacles might you throw in the way?
6. How will you overcome these obstacles?
7. What step can you take in the next week, no matter how small — making a phone call, sending an email, or scheduling in some reflection time — that will move you toward your goal?

When we dare to believe that what we hope for is possible, we can dare to act. As a declaration often inaccurately attributed to Goethe proclaims:

> Until one is committed, there is hesitancy, the chance to draw back. Concerning all acts of initiative (and creation), there is one elementary truth, the ignorance of which kills countless ideas and splendid plans: that the moment one definitely commits oneself, then Providence moves too. All sorts of things occur to help one that would never otherwise have occurred. A whole stream of events issues from the decision, raising in one's favor all manner of unforeseen incidents and meetings and material assistance, which no man could have dreamed would have come his way. Whatever you can do, or dream you can do, begin it. Boldness has genius, power, and magic in it. Begin it now.[7]

CHAPTER ELEVEN

Building Support around You

A crucial factor in any process of change is the level of support it receives. By seeking out encouragement, aid, and good counsel, we create a more favorable context both for our projects and for ourselves. Doing so is especially important when we are facing difficult or hostile conditions. In this chapter we will look at how we can cultivate support at the following levels:

- The personal context of our habits and practices
- The face-to-face context of the people around us
- The cultural context of the society we are part of
- The ecospiritual context of our connectedness with all life

Personal Context: Our Habits and Practices

A good question to start with is "Does the way I live my life support the changes I want to bring about?" Athletes selected for the Olympics are likely to train for years leading up to the event. Their choices about what they eat and drink and when they go to bed are informed by their desire to bring out their best performance. Drawing on advanced methods of sports psychology, they may also use visualization and other practices to strengthen their motivation and sharpen their focus. If we recognize that we are living at a crucial point in human history, when our actions and choices will have consequences lasting thousands of years, is what we do any less important than an Olympic contest?

How we think about the value of what we do is crucial. If we don't see our role as important, we are less likely to take the steps required to function at our best. What helps us recognize our contribution to the Great Turning as significant enough to be taken seriously? The collaborative view of power, explored in chapter 6, certainly makes a difference, since our understanding of emergence helps us appreciate the hidden power of all our choices and actions. The Great Turning requires mass participation, and each of us has a role to play. But do we see our role as important enough to be worth supporting? This is a choice we can make. We can *decide* to give significance to our steps of Active Hope. There are practices that help us do this.

On the last afternoon of a ten-day intensive workshop, Joanna was out walking and met a young monk from the retreat center hosting the event. "Well," he said, "I expect now on your last day you'll be giving people vows." Joanna told him that was not something she did. "Pity," he said. "I find, in my own life, vows so very helpful, because they channel my energy to do what I really want to do."

Continuing on her walk, Joanna looked at her hand and thought that if she *were* to offer vows, they should not number more than the fingers and thumb of one hand. Almost immediately, the following five vows came to her:

> I vow to myself and to each of you:
>
> To commit myself daily to the healing of our world
> and the welfare of all beings.

To live on Earth more lightly and less violently
in the food, products, and energy I consume.

To draw strength and guidance from the living Earth,
the ancestors, the future generations,
and my siblings of all species.

To support others in their work for the world
and to ask for help when I need it.

And to pursue a daily practice
that clarifies my mind, strengthens my heart,
and supports me in observing these vows.

That evening, when Joanna asked the workshop participants what they thought of this idea, "Oh, yes!" was their enthusiastic reply. With the workshop ending, they would soon be scattered far and wide; making these vows to one another and to themselves deepened their sense of mutual belonging as a community. The words "I vow to myself and to each of you" call to mind those we feel are with us as allies.

We need to choose terms that ring true for us. Rather than using the word *vows*, we can, if we prefer, call them "commitments" or "statements of intention." They offer an anchor point to remind us, again and again, of the purposes we hold dear and the behaviors that support us in serving them.

A group's declaration of intent gives tangible form to the shared desires of its members, providing a powerful rallying call. The five vows are an example of this. Now used by many people, they bring a heartening sense of being rooted in a growing global fellowship of intention.

A practice is a habit we have chosen. It is something we have agreed to make a regular feature of our life. Habits develop momentum because the more we repeat them, the more ingrained they become. There are many practices that can support our intention to act for our world. Whether it be meditating, spending time in nature, repeating vows, practicing yoga or tai chi, expressing ourselves creatively, or doing anything else we feel strengthened by, each nourishing activity becomes an ally adding to our context of support.

The fourth vow is to support one another in our work for the world and to ask for help when we need it. This takes us into the second level of context — that involving the people around us.

FACE-TO-FACE CONTEXT: THE PEOPLE AROUND US

Alan, a workshop participant, didn't like to ask for help. "My fear," he said, "is that if I ask for help, people will know I lack the ability to do the task myself. I'd be exposing the fact that I'm not as capable as I'd like them to think I am. That leaves me feeling deficient." In a society that divides people into winners and losers or into those who cope and those who cannot, asking for help can feel like an admission of weakness.

What would happen to a football team if players thought they should be able to cope by themselves and so never passed the ball? The Great Turning is like a team sport; it is a collaborative process. There are times when we need to pass the ball, to call in the assistance of others, and times when we can offer our support too.

When we move from struggling by ourselves to acting as part of a network, we experience the shift in power expressed by the phrase "I can't, we can." Since looking for ways to develop synergy and co-intelligence is part of the Great Turning, the very process of seeking support is a positive step for change. One tool to help identify and amplify the help we receive is the support map. Here is a guide to creating one.

Try This: Creating a Support Map

1. Start by placing your name in the center of a piece of paper. Around it, write the names of people who play a significant role both in your personal and in your work life.
2. Draw arrows to represent flows of support, with the arrows going toward your name for the support you receive and away from you for the support you offer. The thickness of the arrow reflects the strength of the support (see figure 18).
3. Once you've drawn your map, ask yourself how satisfied you are with the flows of support you receive and offer. Do you feel well sustained? If not, how could you invite in more of what you need?

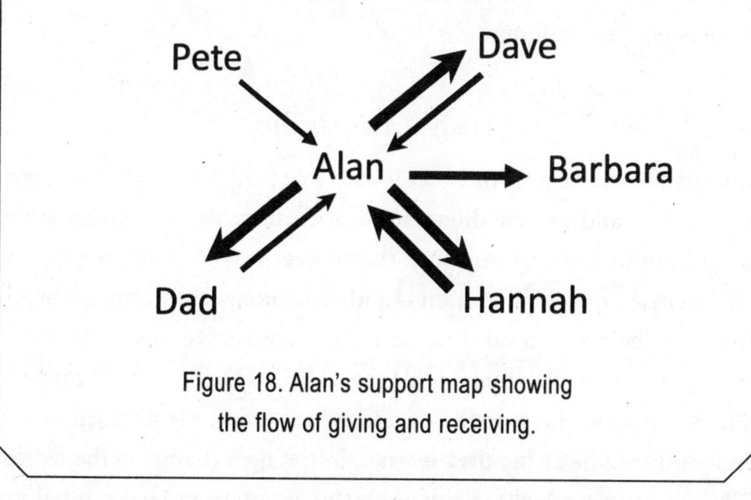

Figure 18. Alan's support map showing the flow of giving and receiving.

When Alan looked at his support map, he was aware of how much better he was at offering support than he was at receiving it. Apart from Hannah, his wife, there were few people he felt supported by, even though he knew his friends Pete and Dave would offer more if he'd let them. Because Alan had been so busy recently, he'd lost track of many old friends. He realized that instead of struggling on alone, he could ask for help, which would give him an opportunity to renew his friendships and make him more effective.

We can also use this mapping process for a specific project, putting the project's name in the center of the paper instead of our own. When you look at the people currently involved and spot gaps or see that you are relying too much on yourself, you can start to think about whom else to recruit. Each project has a life of its own, so it can be interesting to ask, "Through whom does this project want to act?" This question invites us to take a collaborative view. Rather than thinking in terms of "my project" or "their project," we can think of ourselves as part of a team recruited by a vision that wants to happen.

When a deeper purpose acts through people, a special kind of bond can arise between them. A wonderful way to experience this is in a study-action group. For both Joanna and Chris, such groups have been life changing.

Study-Action Groups

On Joanna's return from Tibet in late 1987, she felt a renewed commitment to address the dilemma of nuclear waste. Recognizing the need both to learn more about the subject and to draw in support, she invited a dozen or so friends and acquaintances to join her in addressing the problem of the disposal of radioactive materials. From study groups in the 1970s with the Movement for a New Society, she recognized the benefits of adding an action element, in which the group applies what they learn to initiating a change in the world. This shared practical focus heightens energy, trust, and mutual respect in the group.

The group she formed met monthly, and its work had three main strands, which they called "the three *S*'s": study, strategy, and mutual support. For the study strand, members took turns researching and teaching the others about particular areas, such as the physics of radiation, its biological effects, and the amounts of waste created by nuclear power and weapons production. The strategy side involved taking practical steps based on what they were finding out together. They gave talks in schools, organized petitions, and produced written materials.

Much of the subject matter was deeply disturbing. There were times when they needed to express the grief and anger they felt. That's why the third strand, mutual support, was so important: it made room for the group's emotional responses and helped sustain the members' motivation and courage to keep going.

During the four or so years the group met, they developed a training for themselves that equipped them to offer testimony at public hearings, conduct public education events, produce publications, and create a code of ethics on the care of radioactive materials. The three strands of study, strategy, and mutual support worked together to make their time with the group a treasured experience.

Inspired by Joanna's experience, Chris and friends set up a study-action group about the Great Turning. The first step was to invite interested people to hear more about the idea and say what they were able to commit to. Everyone who came was keen to take part, and the group agreed to meet monthly for six sessions, then to review. Keeping the size to just a dozen people to retain a sense of intimacy, they embarked on a journey guided by the questions "What is the Great Turning?" and "How can we take part in it?" A deeply nourishing feeling of fellowship grew over the eighteen months the group lasted before it came to a natural ending point.

Study-action groups are attractive because they can happen in our homes, they don't require a high level of expertise to set up, and they are enjoyable. The key ingredients needed are a few people sharing a desire to learn more about an issue plus plenty of curiosity, willingness, tasty snacks, and a place to meet.

The study-action model we've described is only one of many forms a support group can take. If we find a book inspiring, another option is to gather a group to read and discuss it together. Doing so gives us an opportunity to read at a deeper level, helping us to digest new information and apply it in our lives.

Over the past decade, we've been pleased to see groups in many countries using this book as a guide to a group journey that nourishes our capacity to show up and play our part. Active Hope book groups, known also as Circles of Active Hope, might either be set up and facilitated by persons experienced in this work or be self-organized and peer led. Sometimes they've involved just a few friends or family members reading the book together and also engaging in the practices. We've developed a free online resource to support people in setting up groups like this and also have a free video-based online course that can be used alongside, or instead of, the book.[1]

CULTURAL CONTEXT

In their groundbreaking study involving more than a hundred thousand survey responses and hundreds of focus groups, social psychologists Paul Ray and Sherry Ruth Anderson describe the growth of a new subculture committed to ecological values, social justice, and holistic perspectives. Extrapolating their findings, they estimate that in the United States, tens of millions of these "cultural creatives" are leaving behind the old story of Business as Usual and creating something new: "When tens of millions of people make such choices in the space of a few decades, we are witnessing not simply a mass of personal departures but an exodus at the level of culture itself."[2]

Although this shift has now grown to involve hundreds of millions of people internationally, it is not always easy to see, and as a result individuals often feel alone in their concern for the world. When we're reading, watching, or listening to corporate-controlled media or walking down a busy shopping street, we might struggle to find evidence that others have noticed the planetary emergency we face. How can we cultivate a more supportive context at the level of our culture and society?

A major factor influencing our decisions is what we believe others around us are doing. Research shows that people are more likely to reduce their energy consumption when they know their neighbors are doing so.[3] Each of us has a reference group of those we compare ourselves to in determining what is normal or appropriate behavior. We also feature in other people's reference groups, so when they see us taking steps to live more sustainably, they are more likely to take these steps too.

The recognition of the power of example at a local level has led to a kind of community organizing that now involves hundreds of thousands of people. Developed through decades of action research, the Cool Block campaign on carbon reduction brings together people from the same street or block in small "eco-teams" to create measurable reductions in their carbon footprint.[4] More recently, in 2021, the concept has been expanded from Cool Blocks to Cool Cities, with three California cities — Los Angeles, Irvine, and Petaluma — committing to an ambitious pathway to carbon neutrality within ten years.

One finding of the Cool Block program is that in spite of initial skepticism about people's willingness to work with their neighbors on cocreating sustainable lifestyles, once the process got started, it gained a momentum of its own. As David Gershon, who developed this approach, describes:

> I started with the point of view of thinking people won't want to disturb their neighbors, they'll be rejected. In New York City, where I live, people said, "We don't talk to our neighbors and we're happy about that. We like our individuality." This is not actually the case and it is a paradox. It is not so much that people don't want to know their neighbors, it is that

> they don't know how to connect with the people living next door and build community. As a result, we struggle as isolated and alienated individuals.
>
> The resistance of "I don't know my neighbors, we've never done anything like this, I'm afraid I'll be rejected" is something we've experienced everywhere we've worked.... But once we get people through this using the organizing tools we provide, they come out the other side feeling incredibly excited to have that connection. Again and again, people say that what they like most about the program is getting to know their neighbors.[5]

When volunteers were recruited, trained, and supported to knock on the doors of their neighbors' homes — inviting them to join small local groups making lifestyle changes to reduce their carbon footprints — the participation rate was more than 40 percent. The level of concern, as well as the willingness to become involved, was much higher than was anticipated. The desire to take part in the healing of our world seems to be just below the surface, waiting for an opportunity and outlet for expression.

Whenever we bring the desire for the world's healing out into the open, whether through our individual actions or through the groups we are part of, we help others do this too. The power of example is contagious. This is how cultures change.

THE ECOSPIRITUAL CONTEXT: OUR CONNECTEDNESS WITH ALL LIFE

Recent research gives strong support to something we know from our own experience: contact with the natural environment can be powerfully restorative to our well-being.[6] Prisoners who can look outdoors from their cells get sick less often, and patients in hospitals recover more quickly when their view is of greenery rather than concrete. In developing a context that supports us in working for our Earth, we need to include contact with the natural world.

There is something more fundamental here than the feel-good factor of pleasant scenery. Particularly for those living in cities, it is easy to lose touch with the biological reality that we humans arise out of nature and are part of it. While the Haudenosaunee and other indigenous peoples recognize that our very survival depends on the healthy functioning of the natural world, it is only recently that this foundation of well-being has become more widely accepted.

We wouldn't have the oxygen we breathe if it weren't for plant life and plankton. We wouldn't have the food we eat if it weren't for the rich living matrix of soil, plants, pollinating insects, and other forms of life. When we carry within us a deep appreciation of how our life is sustained by other living beings, we strengthen our desire to give back.

With the first three levels of context — the practices, the people, and the cultures we're part of — we're referring to tangible behaviors and entities we can point to and that other people can observe. With this fourth level of context, we're describing an experience of connection that leaves us feeling held and supported wherever we happen to be, even when there's nobody else around. We call this fourth context *ecospiritual* because it is about our felt relationship to our planet itself as a source of life; it is as intimate as the rhythm of breathing in and breathing out.

Each of us has an inner map marking realms we regard as supremely important or sacred, purposes we deem essential, and sources we trust for renewal and guidance. In the map of reality that accompanies the Business as Usual story, money is what matters most; anything that gets in the way of short-term profit becomes an unfortunate casualty. On this map, the route to gaining support is through acquiring the money needed to pay for it. This map is taking us over a cliff.

The third dimension of the Great Turning involves a shift in consciousness. We can think of this shift as changing our map to one that puts the healing of our world at the very center of things. On this map, our community is all of life. So we know the trees, the insects, and birds, as well as the clouds, winds, and air, as our kith and kin, as our extended family, as parts of our larger ecological self. With this map of reality, our ecospiritual context is populated with both supportive and destructive forces, with the rain and sun we need, as well as with endangering hurricanes, droughts, and wildfires.

The desire that life continue is far larger than we are. When the choices we make are guided by this desire, we can imagine around us the cheering of all who share it: the ancestors, the future beings, the natural world itself. When we're feeling alone, discouraged, or desperate, we can turn to any of these for support.

If we're not accustomed to seeking solace from the natural world, a good place to start is by casting our mind back to our fondest memories of times spent in nature. Were there any places where we felt especially at peace or that we particularly loved to visit? Drawing on our memories and imaginations, we can mentally return to these places to recall what we received and, in reliving the experience, receive it again. It is good to find a special place in nature where we can go for repeated contact. We can think of this as similar to visiting an old friend or mentor. Can a place really speak to us? Try the following practice and see what happens.

TRY THIS: FINDING A LISTENING POST IN NATURE

Is there a place where you feel particularly connected to the web of life? It can be either somewhere you go physically or somewhere in your imagination. Each time you go there, make yourself comfortable. Think of yourself plugging in to a root system that can draw up insights and inspiration as well as other nutrients. To receive guidance, all you need to do is ask for it — and then listen.

CHAPTER TWELVE

Maintaining Energy and Motivation

When we catch the spark of heartfelt activism for our world, that inner fire can be a remarkable source of energy. However, it also brings with it the risk of burnout. How can we remain fired up for any length of time without being driven to exhaustion? In this chapter we look at how to keep motivation and energy fresh by taking on the challenge of actively maintaining them. Putting personal sustainability right at the heart of what we do includes considering ways to make our expression of Active Hope more enjoyable and engaging.

With our world in crisis, it might seem a bit indulgent to be considering our own enjoyment and enthusiasm. With so many pressing issues to address, shouldn't we brush aside concerns about personal gratification? Yet making what we do rewarding has a strategic value, and it goes further than just preventing burnout. Although many millions of people are already involved in the Great Turning, this planetary response needs to grow. The attractiveness of participation increases when it is recognized as a path to greater aliveness and a more satisfying way of life. Here are six strategies that help:

- Draw out change talk from ourselves and others.
- Recognize enthusiasm as a valuable renewable resource.
- Broaden our definition of activism.

- Follow the inner compass of our deep gladness.
- Redefine what it means to have a good life.
- See success with new eyes and savor it.

Let's look at each of these in turn.

Draw Out Change Talk from Ourselves and Others

Imagine yourself in conversation with someone you'd like to support and encourage to make a change. If good reasons for doing things differently could come out of just one mouth, yours or theirs, whose would you choose? If you'd prefer motives for change to be voiced by the person you're supporting, then you've understood a central principle proven to help people change. Motivational Interviewing is a leading approach in health psychology because it has been shown to work.[1] At its core is the skillful practice of drawing out "change talk" from the person being supported.

William Miller and Steve Rollnick, the two psychologists who developed Motivational Interviewing, define change talk as "self-expressed language that is an argument for change."[2] Research shows that when people make their own argument for change by describing why they want it and why it is important to them, as well as what actions they'll take and the commitment they have to doing so, they're more likely to follow through in making that change.[3]

If the person we want to support is ourselves, we can ask ourselves questions or give ourselves prompts that draw out our own change talk. Many of the exercises we share throughout this book serve this purpose: they invite us to express our motivations and commitments, as well as to describe how we might act on these and to identify what resources might support our actions. With each journey round the spiral, we reinforce our motivation by hearing from ourselves, voicing our concerns, and talking ourselves through the steps we'll take in response. When we share these exercises in company, the role we play as witnesses and allies strengthens the impact of the change talk we hear.

Recognize Enthusiasm as a Valuable Renewable Resource

In responding to the problems of our world, we do need to stretch ourselves and face adversities. The problem is that stretching too far and for too long brings with it the risk of burnout. This state of physical and emotional exhaustion is a form of the overshoot and collapse we described earlier; it is caused by long-term exposure to high levels of stress along with insufficient time and nourishment for renewal. Jenny, who'd worked for many years in a campaigning organization, described what burnout felt like: "I used to love the work I did. I used to enjoy giving talks and engaging with people. But I reached the point where I felt sick of everything. I had worked too hard for too long, and I didn't have anything left to give."

When athletes want to improve their performance, they expect to push beyond their comfort zones. Through the well-established principle of interval training, they alternate periods of intense effort with pauses for renewal. A similar principle applies in yoga when one briefly extends a stretch beyond what is comfortable and then draws back. When we keep pushing, we risk harming ourselves. To develop forms of activism we can sustain for decades, we need to address our requirements for renewal. If we are to create an approach to activism that we will want to stick with over a lifetime and that others are drawn to as well, then we need to look at what feeds our enthusiasm.

Sustainable agriculture reveals how valuable a resource healthy soil is. Finding ways to nourish, renew, and restore soil is a key to long-term productivity. There is a parallel here with enthusiasm: if we see it as valuable, then we become more interested in how we can nourish and restore this precious renewable resource.

An image of a boat floating on water provides a starting point for mapping out the factors influencing our ability and willingness to keep going (see figure 19). The water level represents our inner reserves of energy and enthusiasm, while hitting a problem like burnout is like crashing into a rock. Draining factors are shown as downward arrows that lower the water level and increase our likelihood of hitting rocks. Nourishing factors that replenish and strengthen us are mapped as arrows raising the water level. For example, when we feel we're making progress with our projects, our morale is boosted, and our water level rises. Each time we feel discouraged, whether by arguments, setbacks, or the experience of powerlessness, the water level drops. So to strengthen our resilience, we need to pay attention to all the factors that sustain us.

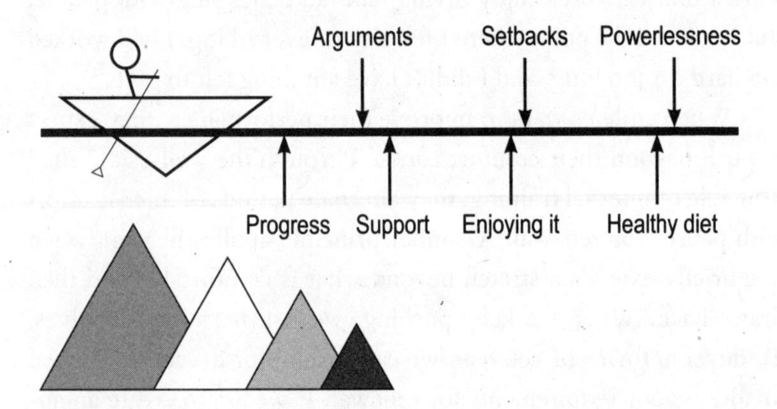

Figure 19. Mapping factors that influence our energy and enthusiasm.

If we experience too much discouragement and too few upward arrows, then we are in danger of reaching such a low ebb that we may wonder what the point of it all is and think about giving up. When we lose the will, energy, and enthusiasm to continue, we're hitting the rock of burnout.

With the worsening condition of our world, the slow pace of progress, and the powerful resistance to much-needed change, protecting our enthusiasm has become especially important. Building support around us, as discussed in the previous chapter, plays a central role here, but it is only one of many potential upward arrows. Once we recognize the value of enthusiasm, we start searching for ways to make what we do more satisfying. The following open sentences invite this exploration.

Try This: Open Sentences for Maintaining Energy and Enthusiasm

These open sentences can be used when journaling, in conversation with friends, or within a group. They work well in conjunction with the water-level mapping process described above.

- Things that drain, demoralize, or exhaust me include...
- What nourishes and energizes me is...
- The times I'm most enthusiastic are when...

What might happen if we applied open sentences like these when planning meetings or events? Many public meetings follow the old model of having an active speaker at the front of the room and a passive audience sitting in rows, with little, if any, interaction. Sometimes these meetings can be informative and inspiring; but they can also be boring and can induce passivity. The water-level diagram can be used to map out factors that make the difference. How can we push up the water level of enthusiasm in meetings so that people look forward to coming rather than just showing up out of duty?

A sustainability group in Frome, England, had a meeting every month, often with a speaker or a film followed by a discussion. Their meetings became much more popular when they started the evening by sitting down together and eating food they'd brought to share. By creating the time and opportunity to talk with one another, they fed the friendships and sense of community that made them look forward to coming. Conversations while eating have led to a sprouting of collaborative projects and activities that have transformed their local community.

Broaden Our Definition of Activism

What is the opposite of an activist? Is it someone who is passive? If so, it seems strange that the term *activist* should be reserved for just a few of us rather than being an identity we all take pride in or aspire to. The practice of Active Hope involves being an activist for what we hope for in the world. We're using the term *activist* here to mean anyone who is active for a purpose bigger than personal gain.

The three dimensions of the Great Turning offer a structure that extends the range of activism beyond the important work of campaigning and protest. Whenever we act from bodhichitta, the desire that all life be well, we are being activists. This includes all endeavors to build a sustainable culture, along with everything that promotes the concurrent shift in consciousness and perception. Having a larger map of activism encourages us to move more freely between these different dimensions, as well as to combine them in ways that empower us. It is possible to overextend ourselves in any of these dimensions, and it can be refreshing to switch our focus when feeling weary.

"Our activism is through our publishing," say Tim and Maddy Harland, who for more than thirty years have produced *Permaculture* magazine. "Our activism is through growing things," say Manu Song and Edi Hamilton, founders of a community agriculture project. Taking part in the Great Turning doesn't have to mean suddenly changing jobs or giving up our other interests. Rather, it means applying our skills, experience, networks, enthusiasm, and temperament to the healing of our world. A participant in an online discussion organized by Transition US said: "The three types of action to achieve the Great Turning offer a very helpful structure. . . . I am by nature better geared to one kind of this work than another, and it is okay to apply myself where I work best."

Follow the Inner Compass of Our Deep Gladness

While there may be periods when we feel ground down or discouraged, there are also times when activism is hugely satisfying, stimulating, and enjoyable. By becoming interested in what makes this so, we can identify what we want to focus on. The flip side of this is that when we feel ourselves going sour inside, experiencing resentment and a loss of our spark, it is worth pausing and reflecting on the choices we can make to restore our enthusiasm. Our degree of enthusiasm can act as a guide, like an inner compass, that helps us steer toward the sort of activity we'll want to continue and deepen for the long term.

We are more effective when acting from our strengths and enthusiasm. That is where the Great Turning can happen through us most powerfully. This is a big shift away from the idea that there is one right way forward we should all follow. Rather, it suggests that each of us needs to find our place of greatest fit. Author and minister Frederick Buechner describes this place of fit as where "your deep gladness and the world's deep hunger meet."[4] When we find this convergence, the Great Turning works through us in a way uniquely our own.

Redefine What It Means to Have a Good Life

The view of a satisfying life presented by glossy magazines and advertisements involves luxury and leisure, illustrated by images of people sunbathing by swimming pools. The scientific research on happiness shows this to be a long way from what really makes life satisfying. Psychologist Mihaly Csikszentmihalyi, in his classic book *Flow: The Psychology of Happiness*, writes: "The best moments usually occur when a person's body or mind is stretched to its limits in a voluntary effort to accomplish something difficult and worthwhile."[5]

Luxury doesn't challenge and stretch us in a way that leads to satisfaction. But activism does in a number of ways. First, when we act in alignment with our deepest values, we experience an inner sense of rightness behind what we do. Second, when we apply ourselves to facing a challenge in a way that absorbs our attention, we are more likely to go into the flow states that psychologists like Csikszentmihalyi have linked so strongly to life satisfaction.

To enter into this kind of flow state, we want to face a challenge difficult enough to absorb us but not so difficult that we feel defeated. When we use our strengths and enthusiasm, we're more likely to enter the nourishing state of absorption where we lose all track of time. This creates a self-reinforcing spiral, where the more we act from our strengths, the more we go into flow, and the more we become absorbed by an activity, the better we get at it. We gain even more satisfaction when virtuous cycles like this contribute to our world.

If we believe the research findings rather than the advertising industry, activism offers a more reliable path to a satisfying life than consumerism. Unfortunately for both our world and our collective mood, this is not yet the dominant view. However, an international movement committed to redefining what it means to have a good life is growing.[6] In Britain, Chris set up the Bristol Happiness Lectures, a series of annual talks that attracted hundreds of people each year. One year the theme was happiness and sustainability. The title of Chris's talk was "How Facing Bad News about the World Can Make You Happier."

At first sight, people may scratch their heads at this title and ask, "Are you serious?" The conventional view is that awareness of world problems is a threat to happiness, and as a result, it is resisted. At issue here are two different understandings of what good mood is based on: one aims for a happy picture, and the other engages in a satisfying process.

The happy-picture approach works along the lines of "If only I had _____ then I'd be happy," happiness here being linked to having the right things. The usual list of desired items includes money, success, and good looks, plus whatever consumer products are most in fashion at the time. If we don't have the right things or look the right way, then our life is not a happy picture. It is easy to feel deficient when measuring ourselves against this manufactured stereotype; our resultant dissatisfaction is a strong driver of consumerism.

One thing you don't find in a happy picture is bad news. Keeping the picture smiley and upbeat involves resisting sources of gloom, so unwelcome features of reality get airbrushed out of view and screened out of conversation. This approach leaves us completely unprepared when we finally get it that we are facing a crisis.

A satisfying process is more like a good story that has both ups and downs. There is no need to hide from bad news here; indeed, it may be the very thing that provokes us to act in a way that makes our life more satisfying. When we rise to a challenge, our strengths are activated, and our sense of purpose is switched on. There is no guarantee we will succeed in bringing about the changes we hope for, but the process of giving our full attention and effort draws out our aliveness. This is what Active Hope is all about: when we live this way, the boredom and emptiness so prevalent in modern society simply disappear.

Someone whose story illustrates the contrast between these two approaches to the good life is John Robbins. At the age of twenty, his life was the American dream come true. His father and uncle had set up the largest ice cream business in history, and Robbins was being groomed to become the next CEO. Something in this happy picture didn't feel right, though. The company's advertising slogan, "We make people happy," didn't ring true to him. While ice cream and smiles might seem to go together, the real consequence of reinforcing this pairing is an increase in obesity and heart disease. What was good for making money wasn't good for people's health. In his book *The New Good Life*, Robbins describes his dilemma: "An ice cream cone never killed anyone, but the more ice cream people eat the more likely they are to develop health problems, and the company naturally wanted to sell as much ice cream as possible."[7]

Robbins felt not only that this obsession with money was wrong but also that it took people away from the things that make life most worthwhile. Describing the pull of a path he found more attractive, he wrote: "I felt called to a different way of life, one whose purpose wasn't focused on making the most money but on making the most difference.... If I were to refuse that call, I might end up rich, but I would surely end up untrue to myself and unhappy. To live against our innermost values can make us sick. It also can lead to disingenuous, counterfeit, or artificial lives."[8]

A truly satisfying process is one we have our heart in; for Robbins, life as a multimillionaire executive didn't offer this. At the age of twenty-one, he turned his back on the vast financial wealth that could have been his. He washed dishes and did other part-time work to support himself through college. In 1969 he and his wife Deo made their home in a tiny one-room log cabin they built on an island in Canada. For the next ten years, they lived a simple life, growing much of their own food and thriving on less than a thousand dollars a year between them. Describing the contrast to the world of money he had grown up in, he wrote: "Both Deo and I wanted to know if something else was possible and might even be more fulfilling... something in our bones told us that humanity was on a collision course with tragedy, and we felt compelled to step out of the rat race so we could more authentically step into our lives."[9]

Some years later, Robbins went on to write the million-copy bestseller *Diet for a New America*, in which he explores the links between food, health, and our environment.[10] The income from book sales didn't deflect John and Deo from their simple life; they'd discovered something on that Canadian island that was both profoundly fulfilling and deeply important to them. Here's how Robbins describes it: "The object of the game is to see how much you can lower your spending while raising your quality of life."[11] It is a game any of us can play, and it has the potential to enrich our lives and transform our world.

If everyone lived as the average North American or European does, we'd need three to five planets like Earth to supply the resources and cope with the rubbish.[12] Because the link between happiness and consumerism has been so powerfully woven into our culture, the idea of giving up or cutting back on what we consume is usually seen as grim and threatening. Yet, as Robbins points out, the real losses come from consumerism. Bit by bit, we are losing our world. We are losing the forests, the fish, the bees; we are wiping out whole species. We are losing the richness of community and much of what makes life meaningful. We are now on the brink of losing the biological support systems we need to survive.

Through the game of learning to live well with less, we stand to gain. We stand to increase our resilience at a time of financial uncertainty, reducing the anxiety over how we'll manage when money is short or loses its value. This game played well leads to happier, more fulfilling lives.

See Success with New Eyes and Savor It

While having our heart in what we do is an essential part of what makes life satisfying, it isn't enough. Repeated failure, frustration, and lack of progress can leave us wondering whether we're wasting our time. It is difficult to stick to a path if we don't see it going anywhere. So the way we understand and experience success affects our willingness to keep going.

The models of success we're likely to have been given generally take us in the wrong direction. In the story of Business as Usual, success is measured in terms of wealth, fame, or position. A company making massive profits is regarded as successful, even when its ways of doing this harm its employees and our world. People are judged as successful simply because they have managed to acquire a vastly larger share of the world's resources than they could ever need — at a time when hundreds of millions starve. It is the very hunger for this type of success that leads us, collectively, to plunder our planet.

With the consciousness shift of the Great Turning, we recognize ourselves as intimately connected with all life, each of us like a cell within a larger body. To call an individual cell "successful" while the larger body sickens or dies is complete nonsense. If we are to survive, could we take the evolutionary step of defining success as that which contributes to the well-being of our larger body, the web of life? Commercial success is easy to count, but how do we count the success of contributing to planetary well-being? Do we experience this success often? And if not, what is getting in the way?

Try This: Reflecting on Success

Taking your definition of success as that which contributes to the well-being of our world, how often do you feel you are succeeding?

We experience success when we reach a goal that is significant to us. But what if our goal is the elimination of poverty or the transition to a low-carbon economy? If the change we want doesn't happen in our lifetime, does that mean we will never experience success? For the encouraging boost we get when we know we're moving forward, we need to find markers of progress we can spot easily and often. What helps here is making the distinction between eventual goals and intermediate ones.

The progressive brainstorming process described in chapter 9 began with longer-range, eventual goals. These are the things we really want to bring about, even if we can't immediately see how. We take one of these eventual goals and list some of the conditions needed to move toward it. So, for example, if our goal is the elimination of poverty, we would need to have in place widespread political will, new taxation policies, redistribution of resources, and the like. Then, taking one of these items, we ask, "What would be needed to accomplish this?" Each stage moves us closer to our present situation. Before long, we're identifying steps that are within our reach, such as eating lower in the food chain or setting up a study-action group on world hunger.

For any goal we choose to pursue, we track back in time to identify intermediate steps. Each time we take an intermediate step, we are succeeding. Instead of rushing on immediately to the next task, we can take a moment to savor our minivictories.

Try This: Savoring Success Every Day

The following open sentence is a useful prompt for this process:

- A recent step I've taken that I feel good about is...

Often we take steps that don't get counted, like the choice of where to place our attention. Just noticing that things are seriously amiss is a step on the journey. If we care enough to want to do something, that is also a significant minivictory. Just to show up with bodhichitta is a success.

In a society that views success in competitive terms, it is usually only those recognized as "winners" who are applauded. We need to learn the skill of encouraging and applauding ourselves. We can reinforce our appreciation of the steps we take by imagining the support of ancestors, future beings, and the more-than-human world. When we develop our receptivity, we will sense them cheering us on. And if we form a study-action group or build support in other ways, we can take time to do this for each other, noticing and appreciating what we're doing well.

When we reflect on past successes, we can ask, "What strengths in me helped me do that?" Naming our strengths makes them more available to us. However, the challenges we face demand of us more commitment, endurance, and courage than we could ever dredge up out of our individual supply. That is why we need to make the essential shift of seeing with new eyes — it takes the process of strength recognition to a new level, that of the larger web of life. Just as we can identify with the suffering of other beings in this web, so too can we identify with their successes and draw on their strengths. There is an ancient Buddhist meditation that helps us do this. It is called the Great Ball of Merit, and it is excellent training for the moral imagination.

Try This: The Great Ball of Merit

Relax and close your eyes, relax into your breathing... Open your awareness to all the beings who share with you this planet time... in this room... this neighborhood... this town. Open to all those in this country and in other lands.

Imagine you can see them in their multitudes, as your awareness widens to include them all.

Opening now to all time, let your awareness encompass all beings who have ever lived... of all species, races and creeds and walks of life, rich and poor, kings and beggars, saints and sinners... Like successive mountain ranges, vast vistas of these fellow beings present themselves to your mind's eye.

Now open to the knowledge that in each of these innumerable lives some act of merit was performed. No matter how stunted or deprived the life, there was at the very least one gesture of kindness, one gift of love, one act of valor or self-sacrifice on the battlefield or in the workplace, hospital, or home... From each of these beings in their endless multitudes arose actions of courage and kindness, of teaching and healing... Let yourself see these manifold and immeasurable acts of merit.

Now imagine that you can sweep together these acts of merit. Sweep them into a pile in front of you. Use your hands... pile them up... pile them into a heap, viewing it with gladness and gratitude. Now pat them into a ball. It is the Great Ball of Merit. Hold it now and weigh it in your hands... Rejoice in it, knowing that no act of goodness is ever lost. It remains ever and always a present resource... a means for the furtherance of life... So now, with jubilation and thanksgiving, you turn that great ball, turn it over... over... into the healing of our world.

The more we practice this meditation, from a Buddhist scripture of the first century CE, the more familiar we become with the process of drawing strengths from outside our narrow selves. Knowing about the Great Ball of Merit can also change the way we think about our own actions. Each time we do something — no matter how small — that is guided by bodhichitta and contributes to our world, we know we are adding to this abundance.

CHAPTER THIRTEEN

Opening to Active Hope

If you hold a tennis ball in the palm of your hand, squeeze it tight, and then let go, the ball quickly springs back into shape. That's how resilience is often thought of, as the capacity to bounce back to a former state after a disturbance. Compare that with squeezing a tomato. Here you get collapse. The toxic false belief in Business as Usual thinking is that we can't change the world. No matter how much we squeeze, there's an assumption that our world system can't be pushed out of shape — that it just springs back like a tennis ball. But the mess we're in is proof this isn't true.

In his study of factors threatening our civilization, Jared Diamond identifies a dozen issues that are "like time bombs."[1] Any of these, including climate change, water scarcity, overconsumption, overpopulation, habitat destruction, loss of topsoil, and rising toxin levels, could trigger the collapse of our society. In combination, their potential impact is even more devastating. Describing his research into the existential threats we face, science writer Julian Cribb said: "I was meeting more and more people who were starting to wonder whether we were entering the 'end game' of human history."[2] The uncertainty we face can be expressed in just four words: Will we make it?

While there might be aspects of our society we'd be pleased to see collapse, the prospect of a wholesale catastrophe ahead is terrifying. When the future seems like a cliff edge we're heading toward, feelings of panic, defeat, and paralysis are understandable. "What's the point? I just feel like giving up" are phrases often spoken or heard as hopes for our lives and our world, like the tomato under pressure, get squished. If you're a young person starting out, it can feel like your future has been stolen.

What helps us face the mess we're in and respond with Active Hope? The story of the Great Turning includes a vital ingredient that is game-changing. It is the unexpected resilience and creative power of life itself. The tomato reminds us that collapse doesn't have to be the end. If it is squeezed in a suitable environment, it is possible to imagine returning a year, a decade, or even a century later and see tomato plants growing. Seeds released by squishing may survive harsh winters or droughts, sprouting into new life time after time. This different kind of resilience is a powerful force of nature. It can be seen in the tender green shoots that bring burned-down forests back to life. It can also be recognized in our own creative power to redesign, adapt, and transform our ways of living, our patterns of organization, and our essential view of life.

In this time of uncertainty, our choices and actions play a role in shaping what happens next. Even if it does turn out to be the endgame for our civilization, there are still different versions of how that might go. If we hold in mind the best that could happen as well as the worst, what can we do to make better scenarios more likely and worse ones less so? The process of turning up to play our part, turning toward our hopes, and turning away from behaviors that make our fears more likely is something we can do from any starting point. We live at a time when the stakes are so high, when so much rests on what we do, that, as Greta Thunberg said, "It's never too late to do as much as we can."[3]

A central question we've been exploring is "What happens through you?" Which story do you hope will grow stronger and happen more through you? If you hope for the Great Turning, the practice of Active Hope involves taking steps that help bring it into being. That includes planting seeds that might sprout into future possibilities, as well as nurturing, developing, and supporting expressions of a life-sustaining culture already here.

There will be times when we feel more open to Active Hope, more engaged and in the flow. We'll also likely experience points when we lose heart, run out of steam, close down. Even then, we can always choose to turn back toward life. In this last chapter, we look at opening as an active process. We focus on three ways of doing this: we call them the Three Acts of Opening.

Like three acts in a play, each builds on the one before. We can also see them in a nonlinear way, as mutually reinforcing steps we can take at any time in any order. They serve as navigational reference points that help us check our bearings. For example, just asking ourselves, "Am I doing this act?" is a way of checking whether we're engaged. We can also see them as options available to us, as responses we can call upon in our practice of Active Hope. The three acts are:

- Act 1: Opening Our Eyes
- Act 2: Opening to Synergy
- Act 3: Opening to Life Acting through Us

Act 1: Opening Our Eyes

The greatest gift we can give our world is the act of paying attention. If we don't have our eyes or other senses open, how can we know what's really happening? How can the situations we face be real to us? If the Great Turning involves turning up to play our part, opening our eyes and our attention is how we begin.

This might seem such a small thing, but when the default setting of mainstream society is to avoid looking at painful realities, it takes courage and determination to sustain the gaze. To see and take in what's actually happening, we need to train ourselves in skills, strengths, and practices that help us keep our eyes open. This will be particularly important as times get more difficult in the years ahead.

We'll need practices like the spiral of the Work That Reconnects. The starting point of gratitude reminds us that opening our eyes doesn't just mean acknowledging what's difficult. It also includes noticing what we love, prompted by that first sentence starter we looked at in chapter 2. The words "I love..." take your eyes and attention on a journey. Our friend and colleague Barbara Ford calls it "wandering the world with grateful eyes." Why not take a moment to try it now? When looking around you, wherever you are, what do you see that you value, appreciate, and/or love?

Noticing what we love can trigger awareness of how much it is threatened. Opening our eyes to our pain for the world is part of giving our attention to reality. If ever you feel squished like a tomato, like you're collapsing in a mess, a useful phrase to reframe this is "This is act 1 — this is opening my eyes."

When we look inward, something else we can access is our imagination. This can open us to a treasure trove of inspiring perspectives and possibilities. A mind experiment that begins with the words "What if..." invites us to consider how things could be different. For example, what if we were somewhere far away in time or space and hearing about life on Earth in this time of extreme danger? We might wish we could be there to play our part. If we wanted a life rich in meaning and purpose, one in which we could really do something that made a difference, this would be a great point in time and space to be alive. Then imagine being able to click your fingers and make it so. *Click*. Here we are.

ACT 2: OPENING TO SYNERGY

Geese can fly more than 70 percent further when traveling in their V-shaped formation than they can if journeying alone. The swirling air in the wake of each bird generates an uplift that helps those close behind it. As it is harder work being out in front, they take turns leading. We can learn from geese and their inspiring demonstration of synergy.

A vow we introduced in chapter 11 was "to support others in their work for the world and to ask for help when I need it." When we step into a support role, like the geese flying together, we become part of a larger formation. Synergy is an expression of power with: it is the creative power of collaboration. So when we look around us for who and/or what to support, we're opening to synergy. Likewise with asking for help — when we do, we're inviting someone else to express power with alongside us.

Synergy happens when different elements act together to achieve something more or different than might otherwise occur. This can include getting different parts of our life to work together in mutually supportive ways. When we bring a new element into our lives, it's not only the element that makes the difference, but also the way we work with it. For example, when we bring a new intention into our lives or recommit to one already there, the way we work with that intention is an opportunity to open to synergy.

When Joanna was in Tibet, she received an important teaching about the power of intention from watching the monks rebuild the great monastery of Khampagar. Once a major center of Tibetan Buddhist culture and learning, it had been almost totally destroyed by the Red Guards during the Chinese Cultural Revolution. After some twenty-seven years, a shift by China to a more relaxed occupation policy allowed reconstruction to begin. This policy, however, could be reversed at any moment, and there was no guarantee that the monastery, if rebuilt, would not be destroyed again. That didn't stop the monks. They faced the uncertainty by bringing to it their intention. They assumed that since you cannot know, you simply proceed. You do what you have to do. You put one stone on top of another and another on top of that. If the stones are knocked down, you begin again, because if you don't, nothing will be built. You persist. In the long run, it is persistence that shapes the future.

The monks working to restore the monastery at Khampagar shared a purpose, direction, and vision. They acted together as part of a larger formation. As with the geese, opening to synergy supported their persistence, helping them to keep going.

ACT 3: OPENING TO LIFE ACTING THROUGH US

Larger wholes act through their parts. You can see that when a good team works well: it acts through its members. If we want to show up the best way we can as a member of a team, an important first step is to think more in terms of "we" and "us" than "I" and "me." If you're part of a football team, the focus shifts from "How do I score?" to "How do we score?" That way you play your part in a larger formation and open to the team acting through you.

The journey we've been taking round the spiral of the Work That Reconnects is about coming back home — coming back to this, our planet home, and to the larger life we are part of. In the shift from being separate people to planet people, we recognize ourselves as part of a larger team of life that can act through us.

The stronger your sense of belonging to a team, a family, or a community, the more you'll feel loyalty and a guiding intention to act for what you belong to. With widening circles of collective identity grows a loyalty to all of life. In chapter 5, we introduced a Buddhist term for this: *bodhichitta*. This is the deep desire to act for the welfare of all beings. When we feel this strongly in our heart, it acts as a foundation stone, something we can count on, whatever else is happening.

Bodhichitta is grounded in our conscious connectedness with all life. In opening to life acting through us, this is our starting point. It is what we build on. Moving around the spiral of the Work That

Reconnects helps us strengthen this connectedness, helps us open to and trust it more. Each time we go around the spiral, we reinforce our bodhichitta. In a time of uncertainty, it can be the one thing we are sure of.

In the Buddhist tradition, bodhichitta is seen as the most precious of all things that we want to treasure and protect. It is pictured as a flame in our heart mind that guides us and shines through our actions. The bodhisattvas, the hero figures of the Buddhist tradition, have such strong bodhichitta that even when, as it is said, they reach the gates of nirvana, having earned the right to disappear into eternal bliss, they turn around every time and choose to come back. They choose to return to samsara, this realm of suffering, because their bodhichitta calls them to serve life on Earth and act for the welfare of all beings.

What if we, like the bodhisattvas, had chosen to return to a life on this world in this time? We don't have to believe in reincarnation to benefit from exploring this different way of thinking about our situation. With this last practice, we offer a process that has deeply affected people in our workshops. We call it Bodhisattva Choices because it involves looking at your life through the eyes of a bodhisattva making the choice to be born and live on this planet in this time.

Introducing the Bodhisattva Choices Practice

Whenever you act for the sake of life on Earth, you express the courageous compassion within you that you can think of as your bodhisattva self. This is part of who you are. It is about showing up as a willing member of a team acting for a purpose much larger than yourself. The bodhisattva archetype is present in all religions and even all social movements. If you have other terms or framings that would help this practice become a better fit for you, please use them.

You're choosing to return not just to *any* life in this planet time, but very specifically to *your* life. There may be roles you are uniquely equipped to play, because of the qualities and life experiences you bring to them.

To consider, even as a thought experiment, that we have chosen our life circumstances can be challenging. Especially where abuse is involved, there is a danger that doing so can be misconstrued as blaming the victim.

Exploring ways to harvest any potential value from our experience of injustice and harmful conditions does not mean we condone these conditions. The psychiatrist Viktor Frankl, subjected to appalling brutality in Nazi concentration camps, engaged in a similar thought experiment when he held in his mind the possible future truth that one day he'd give lectures around the world about the psychology of concentration camps.[4] Finding a context of meaning he could draw strength from offered a source of unexpected resilience.

You can do this process as a personal reflection, writing in a journal or as a letter to yourself. If you are doing it with a friend or in a small group, you can take turns checking in with one another, describing your experience of this process and any thoughts, feelings, or insights it evokes.

TRY THIS: BODHISATTVA CHOICES

The starting point for this process is to imagine yourself poised at the gateway into your present life. You are back before your birth into this present lifetime of yours. For this incarnation, you will have an opportunity to take part in the great changes arising with the dawn of the third millennium; after many decades of growing danger, human civilization is bringing the world to a point of unparalleled peril and promise.

The challenges take many forms — climate catastrophe, the making and using of nuclear weapons, industrial technologies that poison and lay waste to whole ecosystems, billions of people sinking into poverty — but one thing is clear: a quantum leap in consciousness is required if life is to prevail on planet Earth. Hearing this, you decide to renew your commitment to life and reenter the fray by taking birth as a human living on Earth at this time, bringing with you everything you've ever learned about courage and community. This is a major decision. And it is a hard decision, because there is no guarantee that you will remember why you came back and no assurance that you will succeed in your mission. Furthermore, you may well feel alone, because you probably won't even recognize the many other bodhisattvas who, like you, have chosen to be born into this time.

Take a moment to reflect on your willingness to take a human birth in so challenging a planet time. It is your bodhichitta that calls you. Yet this is such a harsh time, and you are aware that you may be born into a life much touched by suffering.

Every human life is by necessity a particular life. You can't take birth as a generic human, but only as a unique human shaped by particular circumstances. So feel yourself stepping into these circumstances now. Imagine that you can understand how they will help prepare you to serve the flourishing of life.

Step into the year of your birth. The timing of your birth allows you to be aware of and affected by particular conditions and events...

Step into the place of your birth. What country did you

choose? Were you born in a town or a city or on the land? Which parts of the Earth's body first greeted your eyes?

Which skin color and ethnicity did you select? And what socioeconomic conditions? Both the privileges and the privations resulting from these choices help prepare you for the work you are coming to do.

Into what faith tradition — or lack thereof — were you born this time? Religious stories and images from childhood — or the very lack of these — influence how you see and search for your purpose.

Now here's an important choice: Which gender or genders did you adopt for this time around? And which sexual orientation?

And as to your parents: Whom did you choose? This might mean your adoptive parents as well as your birth parents. Both the strengths and the weaknesses of your parents, both the loving care you will receive and the hurts you will experience, help prepare you for the work you are coming to do.

Are you an only child, or do you have siblings in this life? The companionship, competition, loneliness, or autonomy associated with that choice will foster the unique blend of strengths you bring to your world.

What physical and/or mental challenges did you choose to take on this time? These differences of ability nourish compassion and deepen your understanding of the difficulties and capacities of others.

Certain strengths and passions characterize this life of yours too. Which mental, physical, and spiritual powers and appetites did you choose to embody for this planet time?

And last, imagining that you can glimpse it for a

moment, what particular mission are you coming to perform?

Each choice relates to your actual life and not to any fantasized alternative to it. What you're doing is seeing these choices from a more encompassing dimension of consciousness. It is as though you're remembering an important aspect of your identity that may have been hidden from view. With this process, you're becoming reacquainted with the bodhisattva within you.

As you become familiar with this exercise, you may wish to add or subtract topics for the bodhisattva's choices. In a follow-up session you can reflect on choices you made in the course of this lifetime, relating, for example, to educational endeavors, spiritual practices, central relationships, and vocational explorations and commitments. All life's experiences, even the harsh and limiting ones, can be seen as ennobling and enriching to our understanding and motivations to serve. After participating in this process, a colleague wrote: "I have been thinking a lot about the Bodhisattva's Choices. I found it very empowering. I consider myself an accountable person. Yet I'd never before systematically reviewed all the major circumstances in my life and celebrated them for bringing me to this time and place."

Finding the Pearl of Active Hope

When Boris Cyrulnik was ten years old, he had to go into hiding. Of Jewish descent and living in France during the Nazi occupation, he had to stay invisible in order to survive. Other members of his family were taken to Auschwitz and killed. Boris's experience of extreme hardship left him with questions about what helps us find strength, what deepens our resilience. Working with abused children, child soldiers in Colombia, and survivors of genocide in Rwanda, he became one of the world's leading psychiatrists addressing children's recovery from trauma. In his book *Resilience*, he writes: "The pearl inside the oyster might be the emblem of resilience. When a grain of sand gets into an oyster and is so irritating that, in order to defend itself, the oyster has to secrete a nacreous substance, the defensive reaction produces a material that is hard, shiny and precious."[5]

With the first act of opening, we open our eyes to both the beauty of our world and the planetary emergency we're facing. When we see what we love, it reminds us of what we act for. When we recognize the danger, it gives us a strong reason to wake up, show up, and play our part.

With the second act of opening, we understand how our acting together generates an impact beyond what we can see. Both offering and receiving support, we find our place within a larger formation.

With the third act of opening, we know that what happens through us is life itself wanting to continue. Looking back at the spiral clock in chapter 8, which represented the flow of life on this planet so far, we see there were past times of mass extinction when the balance hung by a thread. It is an incredible story, marked by episodes of tragic loss and responses of unexpected resilience and creative power. This story continues through us now.

What helps us face the mess we're in and take part in the Great Turning is the knowledge that each of us has something of great value to offer, a priceless role to play. In rising to the challenge of playing our best role, we discover something precious that both enriches our lives and adds to the healing of our world. An oyster, in response to trauma, grows a pearl. We grow, and offer, our response of Active Hope.

Notes

Acknowledgments

1. Thich Nhat Hanh, *The Heart of Understanding: Commentaries on the Prajnaparamita Heart Sutra* (Berkeley, CA: Parallax Press, 1988), 3.

Introduction

1. David Korten, "The Great Turning: From Empire to Earth Community," *YES!*, Summer 2006, 12–17.
2. "Nearly 3 Billion Birds Gone since 1970," Cornell Lab of Ornithology, https://www.birds.cornell.edu/home/bring-birds-back.
3. "Climate Change Widespread, Rapid, and Intensifying — IPCC," Intergovernmental Panel on Climate Change press release, August 9, 2021, https://www.ipcc.ch/2021/08/09/ar6-wg1-20210809-pr.
4. Chi Xu et al., "Future of the Human Climate Niche," *Proceedings of the National Academy of Sciences* 117, no. 21 (May 2020): 11350–55, https://doi.org/10.1073/pnas.1910114117.
5. W. Thiery et al., "Intergenerational Inequities in Exposure to Climate Extremes," *Science* 374, no. 6564 (September 2021): 158–60, https://doi.org/10.1126/science.abi7339.
6. "World Military Spending Rises to Almost $2 Trillion in 2020," Stockholm International Peace Research Institute press release, April 26, 2021, https://www.sipri.org/media/press-release/2021/world-military-spending-rises-almost-2-trillion-2020.
7. For more information, see the Work That Reconnects Network, https://workthatreconnects.org.

8. Rebecca Solnit, *A Paradise Built in Hell: The Extraordinary Communities That Arise in Disaster* (New York: Viking, 2009), 10.
9. Rabindranath Tagore, *Gitanjali* (Brookline, MA: International Pocket Library, 1912), 45.

Chapter One: Three Stories of Our Time

1. *The Ipsos Mori Almanac 2017*, https://www.ipsos.com/sites/default/files/ct/publication/documents/2017-11/ipsos-mori-almanac-2017.pdf.
2. Elizabeth Marks et al., "Young People's Voices on Climate Anxiety, Government Betrayal, and Moral Injury: A Global Phenomenon," *Preprints with "The Lancet,"* September 7, 2021, http://dx.doi.org/10.2139/ssrn.3918955.
3. "Nearly Half the World Lives on Less Than $5.50 a Day," World Bank press release, October 17, 2018, https://www.worldbank.org/en/news/press-release/2018/10/17/nearly-half-the-world-lives-on-less-than-550-a-day.
4. Gernot Laganda, "2021 Is Going to Be a Bad Year for World Hunger," United Nations Food Systems Summit 2021, https://www.un.org/en/food-systems-summit/news/2021-going-be-bad-year-world-hunger.
5. Tim Lenton, Anthony Footitt, and Andrew Dlugolecki, *Major Tipping Points in the Earth's Climate System and Consequences for the Insurance Sector* (Gland, Switzerland: WWF — World Wide Fund for Nature; Munich: Allianz SE, 2009), http://assets.panda.org/downloads/plugin_tp_final_report.pdf.
6. Gordon McGranahan, Deborah Balk, and Bridget Anderson, "The Rising Tide: Assessing the Risks of Climate Change and Human Settlements in Low Elevation Coastal Zones," *Environment and Urbanization* 19 (2007): 17, https://doi.org/10.1177/0956247807076960.
7. "Action against Desertification: Key Facts and Figures," *Organización de las Naciones Unidas para la Alimentación y la Agricultura*, https://www.fao.org/in-action/action-against-desertification/overview/desertification-and-land-degradation/es. See also The World Counts, https://www.theworldcounts.com/challenges/planet-earth/forests-and-deserts/global-land-degradation/story.
8. *Future Earth Risks Perceptions Report 2020*, https://futureearth.org/initiatives/other-initiatives/grp/the-report.

9. Jem Bendell and Rupert Read, *Deep Adaptation: Navigating the Realities of Climate Chaos* (Cambridge, UK: Polity Press, 2021), 23.
10. Assuming large trucks at 25-yard intervals taking 18 tons of waste each, the United States produced 294 million tons of waste in 2018. See "National Overview: Facts and Figures on Materials, Wastes, and Recycling," Environmental Protection Agency, https://www.epa.gov/facts-and-figures-about-materials-waste-and-recycling/national-overview-facts-and-figures-materials.
11. António Guterres, address to UN Assembly, September 22, 2020, in "Covid-19 Dress Rehearsal for World of Challenges to Come, Secretary-General Tells General Assembly," United Nations press release, https://www.un.org/press/en/2020/sgsm20267.doc.htm.
12. US, Brazil, and global total Covid deaths in 2020 taken from Worldometer, https://www.worldometers.info/coronavirus.
13. Sabine van Esland and Ryan O'Hare, "Coronavirus Pandemic Could Have Caused 40 Million Deaths If Left Unchecked," Imperial College London, March 26, 2020, https://www.imperial.ac.uk/news/196496/coronavirus-pandemic-could-have-caused-40.
14. Guterres, address to UN Assembly.
15. Paul Hawken, *Blessed Unrest: How the Largest Movement in the World Came into Being and Why No One Saw It Coming* (New York: Penguin, 2008), 2.
16. Fiona Harvey, "'We've Had So Many Wins': Why the Green Movement Can Overcome Climate Crisis," *Guardian*, October 12, 2020, https://www.theguardian.com/environment/2020/oct/12/climate-crisis-campaigns-pledge-real-change.
17. Jordan Wolman, "Harvard Cracks on Fossil Fuels and a Dam Breaks," *Politico*, September 28, 2021, https://www.politico.com/news/2021/09/28/harvard-fossil-fuels-514483.
18. Quoted in "Graphic: Earth, Discovered," NASA: Global Climate Change, https://climate.nasa.gov/climate_resources/2/graphic-earth-discovered.
19. Ibid.

Chapter Three: Coming from Gratitude

1. Robert A. Emmons, "The Joy of Thanks," *Spirituality & Health*, Winter 2002, https://www.spiritualityhealth.com/articles/2002/01/01/joy-thanks.

2. Emily Polak and Michael McCullough, "Is Gratitude an Alternative to Materialism?," *Journal of Happiness Studies* 7, no. 3 (2006): 343–60, https://doi.org/10.1007/s10902-005-3649-5.
3. R. A. Emmons and M. E. McCullough, "Counting Blessings versus Burdens: An Experimental Investigation of Gratitude and Subjective Well-Being in Daily Life," *Journal of Personality and Social Psychology* 84, no. 2 (2003): 377–89, https://doi.org/10.1037/0022-3514.84.2.377.
4. Polak and McCullough, "Is Gratitude an Alternative," 356.
5. Polak and McCullough, "Is Gratitude an Alternative," 347.
6. Oliver James, *Affluenza* (London: Vermillion, 2007), 28.
7. Oliver James, *Britain on the Couch: Treating a Low-Serotonin Society* (London: Arrow, 1998), 96–110.
8. A. E. Becker et al., "Eating Behaviour and Attitudes Following Prolonged Exposure to Television among Ethnic Fijian Adolescent Girls," *British Journal of Psychiatry* 180 (2002): 509–14, https://doi.org/10.1192/bjp.180.6.509.
9. Jerry Bader, "The Law of Dissatisfaction: How to Motivate Prospects," MRPwebmedia, November 30, 2007, https://www.mrpwebmedia.com/blog/the-law-of-dissatisfaction-how-to-motivate-prospects.
10. Nicole Torres, "Advertising Makes Us Unhappy," *Harvard Business Review*, January–February 2020, https://hbr.org/2020/01/advertising-makes-us-unhappy.
11. Gavin Andrews, Richie Poulton, and Ingmar Skoog, "Lifetime Risk of Depression: Restricted to a Minority or Waiting for Most?," *British Journal of Psychiatry* 187 (2005): 495–96.
12. T. Kasser et al., "Materialistic Values: Their Causes and Consequences," in *Psychology and Consumer Culture: The Struggle for a Good Life in a Materialistic World*, ed. T. Kasser and A. D. Kanner (Washington, DC: American Psychological Association, 2004), 11–28.
13. "A Basic Call to Consciousness: The Hau de no sau nee Address to the Western World, Geneva, Switzerland, Autumn 1977," *Akwesasne Notes*, https://ratical.org/many_worlds/6Nations/6nations1.html#part1.
14. Ibid.
15. "Haudenosaunee Thanksgiving Address: Greetings to the Natural World," National Museum of the American Indian, Smithsonian Institution, https://americanindian.si.edu/environment/pdf/01_02_Thanksgiving_Address.pdf.

16. Ibid.
17. "Message Delivered to the General Assembly of the United Nations by Chief Leon Shenandoah, Special Advisor to the Manitou Foundation Board, on October 24th, 1985," Manitou Foundation, https://www.manitou.org/chief-leon-shenandoah-un-speech-1985.
18. James Lovelock, *Healing Gaia: Practical Medicine for the Planet* (New York: Harmony Books, 1991), 113.
19. Ibid., 22.
20. Ibid., 133.
21. Hamish Gordon and Cat Scott, "Trees Are Much Better at Creating Clouds and Cooling the Climate Than We Thought," *The Conversation*, October 10, 2016, https://theconversation.com/trees-are-much-better-at-creating-clouds-and-cooling-the-climate-than-we-thought-66713.
22. Paying it forward is a powerful concept, beautifully illustrated in the film *Pay It Forward*, directed by Mimi Leder and written by Leslie Dixon, from a book by Catherine Ryan Hyde (Warner Bros., 2000).

Chapter Four: Honoring Our Pain for the World

1. The version used here is based on Joanna's translation of the thirteenth-century poem by Wolfram von Eschenbach.
2. Bibb Latané and John Darley, "Group Inhibition of Bystander Intervention in Emergencies," *Journal of Personality and Social Psychology* 10, no. 3 (1968): 215–21.
3. Quoted in Paul Farhi, "Liberal Media Watchdog: Fox News E-mail Shows Network's Slant on Climate Change," *Washington Post*, December 15, 2010, https://www.washingtonpost.com/wp-dyn/content/article/2010/12/15/AR2010121503181.html.
4. Sutin Wannabovorn, "Thai Meteorology Chief Who Got It Right Is Brought In from the Cold," *Guardian*, January 12, 2005, https://www.theguardian.com/world/2005/jan/12/tsunami2004.thailand.
5. Naomi Oreskes and Erik Conway, *Merchants of Doubt: How a Handful of Scientists Obscured the Truth on Issues from Tobacco Smoke to Global Warming* (New York: Bloomsbury Press, 2010).
6. "Big Oil's Real Agenda on Climate Change: An InfluenceMap Report," March 2019, https://influencemap.org/report/How-Big-Oil-Continues-to-Oppose-the-Paris-Agreement-38212275958aa21196dae3b76220bddc.

7. Marks, "Young People's Voices."
8. Joanna Macy and Molly Young Brown, *Coming Back to Life: The Updated Guide to the Work That Reconnects* (Gabriola Island, BC: New Society Publishers, 2014).
9. For a follow-up study of outcomes in participants, see Chris Johnstone, "Reconnecting with Our World," in *Creative Advances in Groupwork*, ed. Anna Chesner and Herb Hahn (London: Jessica Kingsley, 2002), 186–216.
10. While initially developed in relation to death and dying, the Kübler-Ross model of denial, anger, bargaining, depression, and acceptance is now more widely applied to the process of coming to terms with any kind of loss. See Elisabeth Kübler-Ross and David Kessler, *On Grief and Grieving: Finding the Meaning of Grief through the Five Stages of Loss* (New York: Scribner, 2005).
11. J. William Worden, *Grief Counseling and Grief Therapy: A Handbook for the Mental Health Practitioner*, 4th ed. (New York: Springer Publishing, 2008).
12. Quoted in Macy and Brown, *Coming Back to Life*, 105.
13. Arne Næss, "The Shallow and the Deep, Long-Range Ecology Movement," *Inquiry* 16, nos. 1–4 (1973): 95–100, https://doi.org/10.1080/002017473 08601682.
14. Quoted in *Aulajaaqtut 12, Module 3: Nikanaittuq — Becoming Effective* (Nunavut, Canada: Nunavut Arctic College, 2011), 41, https://www.gov .nu.ca/sites/default/files/tm_12-3_nikanaittuq_-_becoming_effective _web.pdf.

Chapter Five: A Wider Sense of Self

1. Arne Næss, "Self-Realization: An Ecological Approach to Being in the World," in *Thinking Like a Mountain: Toward a Council of All Beings*, ed. John Seed, Joanna Macy, Pat Fleming, and Arne Næss (Kalispell, MT: Heretic Books, 1988), 28–29.
2. Sam Keen, "Hymns to an Unknown God," in *The Sacred Earth: Writers on Nature and Spirit*, ed. Jason Gardner (Novato, CA: New World Library, 1998), 122.
3. Helena Norberg-Hodge, *Ancient Futures: Learning from Ladakh*, 2nd ed. (London: Rider, 2000), 83.
4. Carl Jung, "Spirit and Life," in *The Collected Works of C. G. Jung*, vol. 8, *The Structure and Dynamics of the Psyche* (London: Routledge & Kegan Paul, 1960), 337.

5. Næss, "Self-Realization," 28.
6. Marilynn Brewer, "The Social Self: On Being the Same and Different at the Same Time," *Personality and Social Psychology Bulletin* 17, no. 5 (1991): 476, https://doi.org/10.1177/0146167291175001.
7. Described by Næss in "Self-Realization," 28.
8. Andrews, Poulton, and Skoog, "Lifetime Risk of Depression."
9. Næss, "Self-Realization," 20.
10. Quoted in Joanna Macy, *World as Lover, World as Self* (Berkeley, CA: Parallax Press, 2021), 135.
11. Harvey Sindima, "Community of Life: Ecological Theology in African Perspective," in *Liberating Life: Contemporary Approaches in Ecological Theology*, ed. Charles Birch, William Eaken, and Jay B. McDaniel (New York: Orbis Books, 1990), 137–47, reprinted in Religion Online, https://www.religion-online.org/article/community-of-life-ecological-theology-in-african-perspective.
12. Lynn Margulis and Dorion Sagan, *Microcosmos: Four Billion Years of Evolution from Our Microbial Ancestors* (Berkeley and Los Angeles: University of California Press, 1997), 29.
13. According to the World Health Organization's *Global Status Report on Road Safety 2018*, there are 1.35 million deaths a year from road traffic accidents: https://www.who.int/violence_injury_prevention/road_safety_status/2018/en. In contrast, 250,000 people die yearly from firearms: "Gun Deaths by Country 2022," World Population Review, https://worldpopulationreview.com/country-rankings/gun-deaths-by-country.
14. Karen Kissane, "Black Saturday Fires Equal to 1500 Atomic Bombs: Expert," *Sydney Morning Herald*, May 22, 2009, https://www.smh.com.au/national/black-saturday-fires-equal-to-1500-atomic-bombs-expert-20090521-bh79.html.
15. "List of Major Bushfires in Australia," Wikipedia, lasted edited February 4, 2022, https://en.wikipedia.org/wiki/List_of_major_bushfires_in_Australia.

Chapter Six: A Different Kind of Power

1. "World Troubles Affect Parenthood," BBC News, October 8, 2007, http://news.bbc.co.uk/1/hi/uk/7033102.stm.

2. Joe Litobarski, "We Are Powerless to Stop Climate Change," *Th!nk about It*, November 14, 2009.
3. "World Military Spending," Stockholm International Peace Research Institute.
4. David Laborde et al., *Ending Hunger: What Would It Cost?* (Winnipeg: International Institute for Sustainable Development, 2016), https://www.iisd.org/system/files/publications/ending-hunger-what-would-it-cost.pdf.
5. Nelson Mandela, *Long Walk to Freedom: The Autobiography of Nelson Mandela* (London: Abacus, 1995), 625.
6. D. H. Lawrence, "The Third Thing," *Pansies: Poems* (New York: Knopf, 1929), 133.
7. Mandela, *Long Walk*, 438.
8. Joanna Macy, "Grace and the Great Turning," Co-Intelligence Institute, http://www.co-intelligence.org/Grace_Macy.html.
9. Quoted in Joanna Macy, *Despair and Personal Power in the Nuclear Age* (Gabriola Island, BC: New Society Publishers, 1983), 134. Adapted from T. H. White, *The Sword in the Stone* (New York: Putnam, 1939).

CHAPTER SEVEN: A RICHER EXPERIENCE OF COMMUNITY

1. M. Scott Peck, *The Different Drum: Community Making and Peace* (London: Arrow, 1990), 27.
2. Miller McPherson, Lynn Smith-Lovin, and Matthew Brashears, "Social Isolation in America," *American Sociological Review* 71, no. 3 (June 2006): 353–75.
3. "Cigna Takes Action to Combat the Rise of Loneliness and Improve Mental Wellness in America," Cigna press release, January 23, 2020, https://newsroom.cigna.com/cigna-takes-action-to-combat-the-rise-of-loneliness-and-improve-mental-wellness-in-america.
4. Simon Szreter and Michael Woolcock, "Health by Association? Social Capital, Social Theory, and the Political Economy of Public Health," *International Journal of Epidemiology* 33, no. 4 (August 2004): 650–67, https://doi.org/10.1093/ije/dyh013.
5. Solnit, *Paradise Built in Hell*, 3–4.
6. Ibid., 5.
7. While this well-known quote is widely attributed to Margaret Mead, it

doesn't appear in any of her published writing, and we've found no record of when she said it.

8. Doris Haddock, in "Fear Is a Humbug," *New Hampshire*, December 1, 2008, also at https://www.nhmagazine.com/fear-is-a-humbug.
9. Joanna Macy, *Dharma and Development: Religion as Resource in the Sarvodaya Self-Help Movement*, rev. ed. (Sterling, VA: Kumarian Press, 1991).
10. "US Hate Crime Highest in More Than a Decade — FBI," BBC News, November 17, 2020, https://www.bbc.co.uk/news/world-us-canada-54968498.
11. António Guterres, "We Must Act Now to Strengthen the Immunity of Our Societies against the Virus of Hate," United Nations, https://www.un.org/en/coronavirus/we-must-act-now-strengthen-immunity-our-societies-against-virus-hate.
12. Braver Angels, https://braverangels.org.
13. Martin Luther King Jr., "Letter from Birmingham Jail," April 16, 1963, https://kinginstitute.stanford.edu/sites/mlk/files/letterfrombirmingham_wwcw_0.pdf.
14. "Too Many Children Dying of Malnutrition," UNICEF USA, June 6, 2013, https://www.unicefusa.org/press/releases/unicef-too-many-children-dying-malnutrition/8259.
15. David Smith and Shelagh Armstrong, *If the World Were a Village* (Toronto: Kids Can Press, 2020).
16. Chuck Collins, "Global Billionaires See $5.5 Trillion Pandemic Wealth Surge," Institute for Policy Studies, August 11, 2021, https://ips-dc.org/global-billionaires-see-5-5-trillion-pandemic-wealth-surge.
17. "As a Rich-World Covid-Vaccine Glut Looms, Poor Countries Miss Out," *Economist*, September 4, 2021, https://www.economist.com/international/2021/09/04/as-a-rich-world-covid-vaccine-glut-looms-poor-countries-miss-out.
18. Helena Norberg-Hodge, "The Pressure to Modernise," in *The Future of Progress* (Dartington, UK: Green Books, 1992), reprinted on Local Futures website, May 13, 2010, https://www.localfutures.org/the-pressure-to-modernise.
19. Norberg-Hodge, *Ancient Futures*, 45.
20. See the video documentary *Awakening the Skeena*, directed by Andrew Eddy (Eye Film Releasing, 2010); trailer available on Vimeo at https://vimeo.com/40262066.

21. James Lovelock, *The Vanishing Face of Gaia: A Final Warning* (New York: Allen Lane, 2009), 9.
22. Mohawk Community, "Thanksgiving Prayer," http://www.mohawkcommunity.com/images/Thanksgving_prayer_1_.pdf.
23. Helen Briggs, "Wildlife in 'Catastrophic Decline' Due to Human Destruction, Scientists Warn," BBC News, September 10, 2020, https://www.bbc.co.uk/news/science-environment-54091048.
24. Ibid.
25. Duane Elgin, *Promise Ahead: A Vision of Hope and Action for Humanity's Future* (North Yorkshire, UK: Quill, 2001), 28.
26. Pat Fleming and Joanna Macy, "The Council of All Beings," in *Thinking Like a Mountain*, 79–90.

CHAPTER EIGHT: A LARGER VIEW OF TIME

1. "Collapse of Atlantic Cod Stocks Off the East Coast of Newfoundland in 1992," GRID-Arendal, United Nations Environment Programme, https://www.grida.no/resources/6067.
2. Boris Worm, "Averting a Global Fisheries Disaster," *Proceedings of the National Academy of Sciences of the United States of America* 113, no. 18 (May 2016), 4895–97, https://www.pnas.org/content/113/18/4895.
3. "Market Crisis 'Will Happen Again,'" BBC News, September 8, 2009, http://news.bbc.co.uk/1/hi/8244600.stm.
4. "Quant Trading: How Mathematicians Rule the Markets," BBC News, September 26, 2011, http://www.bbc.co.uk/news/business-14631547.
5. Quoted in Jem Bendell and Jonathan Cohen, "World Review," *Journal of Corporate Citizenship*, no. 27 (June 2007): 10.
6. Greg Palast, *The Best Democracy Money Can Buy* (London: Robinson, 2003), 257.
7. National Commission on the BP Deepwater Horizon Oil Spill and Offshore Drilling, *Deep Water: The Gulf Oil Disaster and the Future of Offshore Drilling; Report to the President*, January 2011, 125, https://www.govinfo.gov/content/pkg/GPO-OILCOMMISSION/pdf/GPO-OILCOMMISSION.pdf.
8. McGranahan, Balk, and Anderson, "The Rising Tide."
9. "Lake Karachay," Wikipedia, last edited November 18, 2021, https://en.wikipedia.org/wiki/Lake_Karachay.

10. "Nuclear and Radiation Accidents and Incidents," Wikipedia, last edited February 10, 2022, https://en.wikipedia.org/wiki/Nuclear_and_radiation_accidents_and_incidents.
11. Malidoma Patrice Somé, *Of Water and the Spirit: Ritual, Magic, and Initiation in the Life of an African Shaman* (New York: Penguin, 1994), 9.
12. Several sources were used for this. For an overview, see "Timeline of the Evolutionary History of Life," Wikipedia, last edited February 17, 2022, https://en.wikipedia.org/wiki/Timeline_of_the_evolutionary_history_of_life.
13. Several sources were used for this. For an overview, see "Timeline of Ancient History," Wikipedia, last edited February 7, 2022, https://en.wikipedia.org/wiki/Timeline_of_ancient_history.
14. Jared Diamond, *Collapse: How Societies Choose to Fail or Survive* (New York: Penguin, 2006), 495.
15. This exercise is adapted from the Seventh Generation process described in Macy and Brown, *Coming Back to Life*, 183.

Chapter Nine: Catching an Inspiring Vision

1. The full speech is available on NPR: https://www.npr.org/2010/01/18/122701268/i-have-a-dream-speech-in-its-entirety?t=1639049546882.
2. Elise Boulding, "Turning Walls into Doorways," *Inward Light* 47, no. 101 (Spring 1986).
3. Jill Bolte Taylor, a neuroanatomist who suffered a stroke, combines her personal experience with neuroscience research in describing the different roles of our two cerebral hemispheres. See Jill Bolte Taylor, *My Stroke of Insight: A Brain Scientist's Personal Journey* (New York: Penguin, 2009).
4. Stephen Covey, *The Seven Habits of Highly Effective People* (New York: Simon & Schuster, 1990), 99.
5. Boulding, "Turning Walls."
6. Ibid.
7. Coleman Barks, trans., *The Essential Rumi* (San Francisco: HarperSanFrancisco, 1995).
8. J. Edward Russo and Paul Schoemaker, *Confident Decision Making: How to Make the Right Decision Every Time* (London: Piatkus, 1991), 111.
9. Rob Hopkins, personal communication, September 16, 2021.
10. Alessandro Casale, "Indigenous Dreams: Prophetic Nature, Spirituality,

and Survivance," Indigenous New Hampshire Collaborative Collective, January 25, 2019, https://indigenousnh.com/2019/01/25/indigenous-dreams.
11. Thomas Berry, *The Dream of the Earth* (San Francisco: Sierra Club Books, 1990), 201.
12. Co-Intelligence Institute home page, http://www.co-intelligence.org.
13. Joseph Campbell, *The Hero's Journey: Joseph Campbell on His Life and Work* (Novato, CA: New World Library, 2003), 217.

CHAPTER TEN: DARING TO BELIEVE IT IS POSSIBLE

1. Quoted in Ellen Gibson Wilson, *Thomas Clarkson: A Biography* (New York: Macmillan, 1989), 11.
2. This quote is widely attributed to Nelson Mandela, but there is no historical record of him having said this. See https://quoteinvestigator.com/2016/01/05/done.
3. Chris Johnstone, *Seven Ways to Build Resilience* (London: Robinson, 2019), 18.
4. Russ Harris, *ACT Made Simple*, 2nd ed. (Oakland, CA: New Harbinger Publications, 2019).
5. Described most movingly in Adam Hochschild, *Bury the Chains: The British Struggle to Abolish Slavery* (London: Pan Macmillan, 2006).
6. Joseph Campbell, *The Hero with a Thousand Faces* (1949; repr., Fontana, 1993), 92.
7. The quote is originally from *The Scottish Himalayan Expedition*, a 1951 book by the Scottish mountaineer W. H. Murray. Murray attributes the two final lines to Goethe, but although Goethe expressed a similar idea in *Faust*, he never wrote these exact lines. See Hyde Flippo, "A Well-Known Quote Attributed to Goethe May Not Actually Be His," ThoughtCo., June 24, 2019, https://www.thoughtco.com/goethe-quote-may-not-be-his-4070881.

CHAPTER ELEVEN: BUILDING SUPPORT AROUND YOU

1. "Study Guide & Resources for Active Hope Book Groups," ActiveHope.Training CIC, https://www.activehope.info/book-groups.
2. Paul Ray and Sherry Ruth Anderson, *The Cultural Creatives: How 50 Million People Are Changing the World* (New York: Harmony Books, 2000), 44.

3. P. W. Schultz et al., "The Constructive, Destructive, and Reconstructive Power of Social Norms," *Psychological Science* 18 (2007): 429–34.
4. David Gershon, *Social Change 2.0: A Blueprint for Reinventing Our World* (White River Junction, VT: Chelsea Green, 2009).
5. Interviewed by Chris Johnstone, "Changing the World from Your Neighbourhood," *Permaculture*, March 7, 2011, https://www.permaculture.co.uk/articles/changing-world-your-neighbourhood.
6. Cecily Maller et al., "Healthy Nature, Healthy People: 'Contact with Nature' as an Upstream Health Promotion Intervention for Populations," *Health Promotion International* 21, no. 1 (December 2005): 45–54, https://doi.org/10.1093/heapro/dai032.

Chapter Twelve: Maintaining Energy and Motivation

1. B. Lundahl and B. Burke, "The Effectiveness and Applicability of Motivational Interviewing: A Practice-Friendly Review of Four Meta-Analyses," *Journal of Clinical Psychology: In Session* 65, no. 11 (2009): 1232–45.
2. William Miller and Stephen Rollnick, *Motivational Interviewing*, 3rd ed. (New York: Guilford Press, 2012).
3. P. C. Amrhein et al., "Client Commitment Language during Motivational Interviewing Predicts Drug Use Outcomes," *Journal of Consulting and Clinical Psychology* 71 (2003): 862–78.
4. Frederick Buechner, "July 18, 2017: Vocation," https://www.frederickbuechner.com/quote-of-the-day/2017/7/18/vocation.
5. Mihaly Csikszentmihalyi, *Flow: The Psychology of Happiness* (London: Rider, 1992), 3.
6. See, for example, Action for Happiness, https://www.actionforhappiness.org.
7. John Robbins, *The New Good Life: Living Better Than Ever in an Age of Less* (New York: Ballantine, 2010), 5.
8. Ibid., 10.
9. Ibid.
10. John Robbins, *Diet for a New America*, 2nd ed. (Walpole, NH: Stillpoint, 1998).
11. Robbins, *New Good Life*, xv.
12. For a helpful handout with more information, see the Global Footprint

Network media backgrounder, https://www.footprintnetwork.org/content/images/uploads/Media_Backgrounder_GFN.pdf.

CHAPTER THIRTEEN: OPENING TO ACTIVE HOPE

1. Diamond, *Collapse*, 498.
2. David Hulme, "Interview: Humanity's End Game," *Vision*, Winter 2018, https://www.vision.org/interview-julian-cribb-interview-human-survival-8209.
3. "Greta Thunberg: 'It's Never Too Late to Do as Much as We Can,'" BBC News, October 31, 2021, https://www.bbc.co.uk/news/av/science-environment-59110260.
4. Viktor Frankl, *Man's Search for Meaning* (London: Pocket Books, 1959).
5. Boris Cyrulnik, *Resilience: How Your Inner Strength Can Set You Free from the Past* (New York: Penguin, 2009), 286.

Resources

If you've been touched by the approach we share and would like to explore further, here are some resources for learning more. We've included ways of finding out about workshops and an international network of facilitators, information about the Work That Reconnects and Active Hope, as well as ways to train to share this work with others.

https://workthatreconnects.org

The Work That Reconnects Network website is a go-to source, with a wealth of information about the ideas and practices of this approach, plus an events listing of workshops and courses, a newsletter, a facilitator directory, and a resources section.

https://vimeo.com/channels/workthatreconnects

The Vimeo channel of the Work That Reconnects Network offers all the videos (more than three hours' worth) from Joanna Macy's DVD set *The Work That Reconnects*. The videos are a great introduction for those interested in facilitation. The channel also includes webinar recordings from leading practitioners on a range of themes linked to the Work That Reconnects.

https://www.activehope.info

This is a go-to source for information about Active Hope, with details of the different language editions, an introduction to some of the key themes, video demonstrations of exercises, webinar recordings, interviews, and an entry point to an online resource for people wanting to set up and lead Active Hope book groups.

https://activehope.training

This website offers a free video-based online course guiding participants round the spiral of the Work That Reconnects, with a series of seven practices suitable for personal use, partnered conversations, or group work. Survey results show people completing the course found it significantly strengthened their motivation to act and their belief they could make a difference, leaving them feeling less overwhelmed, more personally nourished, and more connected with life.

https://www.joannamacy.net

Joanna Macy's website, with a wealth of information about her work and thought, as well as a listing (with links where available) of her books, articles, interviews, podcasts, and videos.

https://chrisjohnstone.info

Chris Johnstone's website, with information about his work and key themes, online courses, books, articles, videos, and music.

https://localcircles.org

In this crisis moment, we're all helped by company. The Resilience Circle Network website is a rich source of information about how to set up and guide resilience circles (small groups of ten to twenty) that bring people together to foster community, mutual support, and social action.

Coming Back to Life: The Updated Guide to the Work That Reconnects

The second edition of this book by Joanna Macy and Molly Young Brown (Gabriola Island, BC: New Society Publishers, 2014) is an ideal companion to *Active Hope* for anyone interested in facilitating the Work That Reconnects.

Index

Page numbers in *italics* refer to illustrations.

About the Authors

Joanna Macy, PhD, a scholar of Buddhism, systems theory, and deep ecology, is a respected voice in movements for peace, justice, and ecology. As the root teacher of the Work That Reconnects, she interweaves her scholarship with six decades of activism. The many dimensions of this work are explored in her books and videos.

Hundreds of thousands of people around the world have participated in Joanna's workshops and trainings. Her group methods, known as the Work That Reconnects, have been adopted and adapted yet more widely in activist movements, academic programs, and community organizing. In the face of overwhelming social and ecological crises, this work helps people transform despair and apathy into constructive, collaborative action. It brings a new way of seeing the world as our larger body, freeing us from the assumptions and attitudes that now threaten the continuity of life on Earth.

Macy spent many years traveling in other lands and cultures, engaging movements for social change, and exploring their roots in religious thought and practice. Much of the work emerged in her first extensive teaching tours in the UK, Germany, and Australia. She also led trainings in Russia, countries of the former Soviet Union, Japan, India, Sri Lanka, Italy, France, and Brazil, as well as the United States and Canada. Her students have carried the work to Mexico, Chile, sub-Saharan Africa, China, Poland, the Czech Republic, Kazakhstan, Thailand, Myanmar, and elsewhere around the globe.

Joanna lives in Berkeley, California, near her children and grandchildren. To learn more, visit **joannamacy.net**.

Chris Johnstone is a codirector of ActiveHope.Training, a nonprofit organization offering online courses that help people grow their capacity to make a difference in the world. He is also one of the United Kingdom's leading resilience trainers, having taught resilience skills for more than thirty years and pioneered the role of such training in mental health promotion, in coaching practice, and in relation to addressing concerns about the world.

After training in medicine and a range of psychological therapies, Chris worked for nearly twenty years as an addictions specialist in the UK National Health Service. Moving over the past two decades more into coaching, training, and writing, his work explores what helps us face disturbing situations (whether in our own lives or in the world) and respond in ways that nourish resilience and well-being. He is author of *Find Your Power* (2010) and *Seven Ways to Build Resilience* (2019).

An activist since his teenage years, Chris first encountered Joanna Macy's work while he was a medical student in London. He's been a trainer in the Work That Reconnects for more than three decades, working closely with Joanna on many occasions and running facilitator trainings in the United Kingdom.

Moving to the north of Scotland in 2011, he began developing his training work online. His video-based online courses now reach many thousands of people, with participants in more than sixty countries.

Chris lives with his wife Kirsty in a small Scottish village. He loves growing fruit and has a passion for playing music, particularly with melodic percussion and hammered dulcimer. He has a website at **chrisjohnstone.info.**

NEW WORLD LIBRARY is dedicated to publishing books and other media that inspire and challenge us to improve the quality of our lives and the world.

We are a socially and environmentally aware company. We recognize that we have an ethical responsibility to our readers, our authors, our staff members, and our planet.

We serve our readers by creating the finest publications possible on personal growth, creativity, spirituality, wellness, and other areas of emerging importance. We serve our authors by working with them to produce and promote quality books that reach a wide audience. We serve New World Library employees with generous benefits, significant profit sharing, and constant encouragement to pursue their most expansive dreams.

Whenever possible, we print our books with soy-based ink on 100 percent postconsumer-waste recycled paper. We power our offices with solar energy and contribute to nonprofit organizations working to make the world a better place for us all.

Our products are available wherever books are sold. Visit our website to download our catalog, subscribe to our e-newsletter, read our blog, and link to authors' websites, videos, and podcasts.

customerservice@newworldlibrary.com
Phone: 415-884-2100 or 800-972-6657
Orders: Ext. 110 • Catalog requests: Ext. 110
Fax: 415-884-2199

www.newworldlibrary.com